KNOWLEDGE AND POWER IN A GLOBAL SOCIETY

SAGE FOCUS EDITIONS

1. **POLICE AND SOCIETY,** edited by David H. Bayley
2. **WHY NATIONS ACT,** edited by Maurice A. East, Stephen A. Salmore, and Charles F. Hermann
3. **EVALUATION RESEARCH METHODS,** edited by Leonard Rutman
4. **SOCIAL SCIENTISTS AS ADVOCATES,** edited by George H. Weber and George J. McCall
5. **DYNAMICS OF GROUP DECISIONS,** edited by Hermann Brandstatter, James H. Davis, and Heinz Schuler
6. **NATURAL ORDER,** edited by Barry Barnes and Steven Shapin
7. **CONFLICT AND CONTROL,** edited by Arthur J. Vidich and Ronald M. Glassman
8. **CONTROVERSY,** edited by Dorothy Nelkin
9. **BATTERED WOMEN,** edited by Donna M. Moore
10. **CRIMINOLOGY,** edited by Edward Sagarin
11. **TO AUGUR WELL,** edited by J. David Singer and Michael D. Wallace
12. **IMPROVING EVALUATIONS,** edited by Lois-ellin Datta and Robert Perloff
13. **IMAGES OF INFORMATION,** edited by Jon Wagner
14. **CHURCHES AND POLITICS IN LATIN AMERICA,** edited by Daniel H. Levine
15. **EDUCATIONAL TESTING AND EVALUATION,** edited by Eva L. Baker and Edys S. Quellmalz
16. **IMPROVING POLICY ANALYSIS,** edited by Stuart S. Nagel
17. **POWER STRUCTURE RESEARCH,** edited by G. William Domhoff
18. **AGING AND SOCIETY,** edited by Edgar F. Borgatta and Neil G. McCluskey
19. **CENTRE AND PERIPHERY,** edited by Jean Gottmann
20. **THE ELECTORATE RECONSIDERED,** edited by John C. Pierce and John L. Sullivan
21. **THE BLACK WOMAN,** edited by La Frances Rodgers-Rose
22. **MAKING BUREAUCRACIES WORK,** edited by Carol H. Weiss and Allen H. Barton
23. **RADICAL CRIMINOLOGY,** edited by James A. Inciardi
24. **DUAL-CAREER COUPLES,** edited by Fran Pepitone-Rockwell
25. **POLICY IMPLEMENTATION,** edited by John Brigham and Don W. Brown
26. **CONTEMPORARY BLACK THOUGHT,** edited by Molefi Kete Asante and Abdulai S. Vandi
27. **HISTORY AND CRIME,** edited by James A. Inciardi and Charles E. Faupel
28. **THE FUTURE OF EDUCATION,** edited by Kathryn Cirincione-Coles
29. **IMPROVING SCHOOLS,** edited by Rolf Lehming and Michael Kane
30. **KNOWLEDGE AND POWER IN A GLOBAL SOCIETY,** edited by William M. Evan
31. **BLACK MEN,** edited by Lawrence E. Gary
32. **MAJOR CRIMINAL JUSTICE SYSTEMS,** edited by George F. Cole, Stanislaw Frankowski and Marc G. Gertz
33. **ANALYZING ELECTORAL HISTORY,** edited by Jerome M. Clubb, William H. Flanigan, and Nancy H. Zingale
34. **ASSESSING MARRIAGE,** edited by Erik E. Filsinger and Robert A. Lewis
35. **THE KNOWLEDGE CYCLE,** edited by Robert F. Rich
36. **IMPACTS OF RACISM ON WHITE AMERICANS,** edited by Benjamin P. Bowser and Raymond G. Hunt
37. **WORLD SYSTEM STRUCTURE,** edited by W. Ladd Hollist and James N. Rosenau
38. **WOMEN AND WORLD CHANGE,** edited by Naomi Black and Ann B. Cottrell

KNOWLEDGE and POWER in a GLOBAL SOCIETY

Edited by
William M. EVAN

University of Pennsylvania

SAGE PUBLICATIONS / Beverly Hills / London / New Delhi

For information address:

SAGE Publications, Inc.
275 South Beverly Drive
Beverly Hills, California 90212

SAGE Publications Ltd
28 Banner Street
London EC1Y 8QE, England

Printed in the United States of America

Library of Congress Cataloging in Publication Data

Main entry under title:

Knowledge and power in a global society.

(Sage focus editions ; 30)
Bibliography: p.
Includes indexes.
1. Science--Societies, etc. 2. Technology--Societies, etc. 3. Internationalism. 4. International cooperation. I. Evan, William M. II. Series.
Q10.K6 306'.4 81-8728
ISBN 0-8039-1659-0 AACR2
ISBN 0-8039-1660-4 (pbk.)

FIRST PRINTING

CONTENTS

Preface 7

Glossary of Abbreviations 9

1. Some Dilemmas of Knowledge and Power: An Introduction
   William M. Evan 11

PART I: INTERNATIONAL SCIENTIFIC AND PROFESSIONAL ASSOCIATIONS (ISPAs) AS TRANSNATIONAL ACTORS 27

2. Alternative Models of ISPAs
   Diana Crane 29
3. Varieties of ISPAs: Their Forms and Functions
   Louis Kriesberg 49
   Commentary: *James R. Kurth* 69
   Commentary: *Henry Teune* 73

PART II: ECONOMIC IMPACT OF ISPAs 77

4. Networks of Economists: Their Role in International Monetary Reforms
   Harry G. Johnson 79
5. Project LINK: Policy Implications for the World Economy
   Lawrence R. Klein 91
6. Global Modeling and Decision-Making
   Mihajlo D. Mesarovic and Barry B. Hughes 107
   Commentary: *Robert W. Russell* 121
   Commentary: *Leonid Hurwicz* 127

PART III: POLITICAL IMPACT OF ISPAs 131

7. COSPAR and SCOR: The Political Influence of the Committee on Space Research and the Scientific Committee on Oceanic Research
   Edward Miles 133

8. Stockholm and Vancouver: The Role of ISPAs at UN Conferences
Anne Thompson Feraru 153
9. The Political Significance of Pugwash
Michael J. Pentz and Gillian Slovo 175
10. The Club of Rome and Its Policy Impact
Alexander King 205
Commentary: *Frederick W. Frey* 225
Commentary: *Robert S. Jordan* 229

PART IV: SOCIAL AND CULTURAL IMPACT OF ISPAs 235

11. Do ISPAs Promote Global Integration?
Robert C. Angell 237
12. International Organizations of Teachers: Their Policy Impact
William G. Carr 255
13. Human Rights Organizations
Harry M. Scoble and Laurie S. Wiseberg 273
Commentary: *James Frederick Green* 294
Commentary: *Berhanykun Andemichael* 297
Commentary: *Harold Taylor* 301

Author Index 307

Subject Index 310

About the Authors 317

PREFACE

The gap between the processes and products of science and the processes and products of political decision-making in the world is so wide that it threatens the survival of mankind. Canons of scientific inquiry developed since the seventeenth century govern the activities of scientists the world over, regardless of national and ideological boundaries. The products of these collective and cumulative activities are critically evaluated by members of various communities of scientists that transcend national borders. Only those scientific products that are replicated and validated are incorporated into the common corpus of scientific knowledge. From this body of knowledge a variety of applications emerge, some of which benefit mankind, as in the case of medical and agricultural applications, and some of which threaten mankind, as in the case of military applications.

By contrast, the political decisions made by nation-states are governed by a calculus of "national interests" and power as filtered through the reigning systems of beliefs or ideologies to which the leaders and dominant ruling groups are committed. Since the diverse national interests and ideologies come into conflict with one another, they eventually lead to the development of political, economic and military strategies that result in gains to some nations and losses to others. In the course of this worldwide political "game," which is regulated by relatively few rules accepted by all the players, problems accumulate of a global nature that threaten the survival of nation-states and of the human species as well. Foremost among the global problems are nuclear weapons systems, hunger and starvation, overpopulation, exhaustion of nonrenewable resources, disarray in the world economy, and environmental pollution.

If any of these global problems are to be solved, the "rationality" of pursuing power through the promotion of national interests will have to be replaced with the rationality of pursuing science and applying scientific knowledge on behalf of the human species on this planet, and not in the name of exclusive national, class, racial, or religious interests. Is there any evidence that such a substitution of rationalities will occur? Is there any discernible trend for the application of scientific knowledge to global decision-making? Otherwise put, is there any detectable progress toward "global humanism"? The evidence for such a development is scant indeed. And yet an innovation of potentially historic significance should not be overlooked. It is the growth, since the emergence of modern science, of international scientific and professional associations (ISPAs). These *trans-national* associations have been founded for the purpose of developing and applying scientific knowledge. In the course of pursuing these objectives, they can and do address global problems outside the framework of warring national interests and ideologies. If these organizations increasingly become the medium for infusing scientific knowledge and technology—from a *global humanist perspective*—into the political and economic decision-making process at the international level, there is a chance that some of the worldwide problems may yet be solved in a rational as well as in a humane manner.

To explore the potential influence of ISPAs on global problems and international decision-making, a special conference was recently convened at the University of Pennsylvania. This book reports the proceedings of this unique conference. It is divided into four parts: Part I consists of a discussion of the transnational objectives and activities of international scientific and professional associations (ISPAs); Part II deals with the economic policy influences of ISPAs; Part III with their political impact; and Part IV with their social and cultural impact.

For making possible the convening of this conference, I am indebted to several organizations: Rena and Angelius Anspach Institute for Diplomacy and Foreign Affairs of the University of Pennsylvania, the National Science Foundation, the World Affairs Council of Philadelphia, and the University City Science Center. I should also like to express my gratitude to the following individuals for moral and material support in conducting the conference: Professor Frederick W. Frey, Dr. Randall M. Whaley, Mr. John F. Reichard, Dr. Robert F. Duvall, Ms. Laura B. Farmelo, Professor Arthur I. Bloomfield, Dr. Lalit Aggarwal, Dr. Albert G. Crawford, Dr. Patrick H. Irwin, and Raima Evan. And last but not least, I am deeply indebted to the fifteen authors of the invaluable essays published in this book and to the nine discussants for their provocative commentaries.

Philadelphia, Pennsylvania William M. Evan

GLOSSARY OF ABBREVIATIONS

ACMRR	Advisory Committee for Marine Resources Research
AI	Amnesty International
ASA	Atomic Scientists Association
CEMA	Council for Economic Mutual Assistance
CETEX	Ad Hoc Committee on Contamination by Extra-Terrestrial Exploration (of ICSU)
COSPAR	Committee on Space Research (of ICSU)
ECOSOC	Economic and Social Council of the UN
ESA	Department of Economic and Social Affairs (of UN New York Secretariat)
FAS	Federation of American Scientists
FAO	Food and Agriculture Organization of the UN
FISE	International Federation of Teachers Syndicates
GARP	Global Atmosphere Research Program
GATT	General Agreement on Tariffs and Trade
IACOMS	International Advisory Committee on Marine Sciences
IAEA	International Atomic Energy Agency
IAF	International Astronomical Federation
IAU	International Astronomical Union
IBE	International Bureau of Education
ICAO	International Civil Aviation Organization
ICID	International Commission on Irrigation and Drainage
ICJ	International Court of Justice; International Commission of Jurists
ICPO	International Criminal Police Organization
ICN	International Council of Nurses
ICRC	International Committee of the Red Cross
ICSU	International Council of Scientific Unions
ICUBE	International Christian Union of Business Executives
IEC	International Electrotechnical Commission
IFALPA	International Federation of Airline Pilots Associations
IFFTU	International Federation of Free Teachers Union
IFIAS	International Federation of Institutes of Advanced Study
IGOS	International Governmental Organizations
IGY	International Geophysical Year
IHB	International Hydrographic Bureau
IIC	International Invisible College
ILA	International Law Association
ILHR	International League for Human Rights
ILO	International Labor Organization

IMCO	Intergovernmental Maritime Consultative Organization
IMF	International Monetary Fund
INGOs	International Non-Governmental Organizations
IOC	Intergovernmental Oceanographic Commission
IOE	International Organization of Employers
IOS	International Organization for Standardization
IRCC	International Radio Consultative Committee
ISA	International Sociological Association
ISPAs	International Scientific and Professional Associations
ISSC	International Social Science Council
ISSS	International Society of Soil Sciences
ITU	International Telecommunication Union
IUA	International Union of Architects
IUCAF	Inter-Union Committee on Frequency Allocation
IUCNNR	International Union for Conservation of Nature and Natural Resources
IUNS	International Union of Nutritional Sciences
IUPDEE	International Union of Producers and Distributors of Electrical Energy
IURS	International Union of Radio Sciences
IUSSP	International Union for the Scientific Study of Population Problems
LOS	Law of the Sea
MNCs	Multinational Corporations
NAS/NRC	National Academy of Sciences/National Research Council
NEA	National Education Association
NGOs	Non-Governmental Organizations
NNGOs	National Non-Governmental Organizations
NSPA	National Scientific and Professional Association
OECD	Organization for Economic Cooperation and Development
REMP	Research Group for European Migration Problems
SCIBP	Special Committee on International Biological Programs
SCOPE	Scientific Committee on Problems of the Environment (of ICSU)
SCOR	Scientific Committee on Oceanic Research (of ICSU)
SIPRI	Stockholm International Peace Research Institute
UN	United Nations
UNCHE	United Nations Conference on Human Environment
UNCTAD	United Nations Conference on Trade and Development
UNEP	United Nations Environmental Program
UNESCO	United Nations Educational, Scientific and Cultural Organization
UNICEF	United Children's Fund (United Nations International Children's Emergency Fund)
UNIDO	United Nations Industrial Development Organization
UNITAR	United Nations Institute for Training and Research
WCOTP	World Confederation of Organizations of the Teaching Profession
WEC	World Energy Conference
WFSW	World Federation of Scientific Workers
WHO	World Health Organization
WIM	World Intergrated Model
WMO	World Meteorological Organization

1

SOME DILEMMAS OF KNOWLEDGE AND POWER

An Introduction

William M. Evan

"Knowledge is power," said Francis Bacon, the apostle of modern science. Unlike some other values human beings create that are perishable, scientific knowledge tends to endure and to cumulate, enriching the lives of succeeding generations. This is indeed the premise upon which universities were first founded in the Middle Ages and the reason they have flourished for centuries, whereas many other types of organizations have been ephemeral.

Valuable as scientific and technological knowledge may be, is it powerful? In other words, does knowledge guide the fateful decisions taken at national and international levels that affect the lives of millions of human beings? Does knowledge contribute to the wise use of power? In uttering his aphorism, was Bacon describing reality at the dawn of the seventeenth century or was he projecting a vision of a utopia—in which knowledge *would* be power—as he subsequently did in his *New Atlantis*? These questions point to the uneasy and perplexing relationship between knowledge and power. On the one hand, knowledge may be entirely divorced from power; it may be pursued and developed without any concern for its impact on the exercise of power. On the other hand, knowledge may be cultivated for the express purpose of undergirding the exercise of power.

If, for the purpose of this discussion, we restrict the concept of knowledge to the fruits of science and technology, and the concept of power to public policy decisions of nation-states and international bodies, it is clear that knowledge-producers and power-wielders or decision makers are quite distinct roles: "men of knowledge" are rarely "men of action," except in Plato's *Republic,* which is governed by philosopher-kings. It is equally clear that when knowledge is applied, that is, converted to power, it is never a direct or an automatic process: intervening between knowledge and power are an implicit or an explicit affirmation of values and a commitment to specific interests. Moreover, the values prompting the application of scientific knowledge may be benevolent as well as malevolent, altruistic as well as selfish, humane as well as inhumane. Since the social roles of the knowledge-producers—the scientist, the engineer or the professional, in general—and the decision makers are distinct, there may or may not be consensus among the occupants of these roles regarding the values and interests underlying the utilization of knowledge. And in view of the fact that scientists and technologists rarely participate directly in the decision-making process concerning the utilization of scientific knowledge, do they or should they bear responsibility for the social consequences of the use of their knowledge in such decisions (Lakoff, 1977)?

Yet another complication in the relationship between knowledge and power is that scientific knowledge has its limits: it may often be inadequate or even completely unavailable for the solution of either a particular national or a global problem. Thus, the decision maker may turn in vain for guidance on a policy problem to the scientific and technological community. Finally, the uses of power at national or international levels—for example, for cooperative, competitive, or aggressive purposes—influence the type of knowledge that is developed.

In view of the complexities of the relationship between knowledge and power, how shall we regard the dismaying litany of world problems: the threat of nuclear war, the progressive exhaustion of nonrenewable resources, the rapid growth of population, the increasing gap between rich and poor countries, instability of the world economy, and environmental deterioration? By definition, all these problems are global in scope, transcending national boundaries. If they are ever to be solved, these problems would require the development and application of scientific and technological knowledge. However, the very process of developing and applying knowledge to these problems presupposes a commitment to the values of a "global society."[1] And yet, the probability that the requisite knowledge will be developed and brought to bear on such global problems, *in the name of humanity,* seems at present to be quite low.

One of the explanations for the failure to link knowledge with power in coping with global problems may be found in Bacon's concept of four "idols" in his *Novum Organum*: "idols of the tribe," "idols of the market-place," "idols of the cave," and "idols of the theater" (Fowler, 1881). Today, Bacon's "idols" or "phantoms of the mind" can be freely translated as distortions of reality traceable to a particular ideology or to "conventional wisdom," all too often used as a substitute for systematic scientific inquiry. The widespread commitment to nationalistic values impedes the emergence of transnational values concerned with the survival and welfare of the human species. Another factor in the failure to integrate knowledge with power on behalf of a "global society" is the absence of an infrastructure of scientific and technological research institutes with a mandate to study global problems. Such a deficiency is not unrelated to the failure of the international system to develop effective international legal remedies—for example, by unequivocally granting jurisdiction to the International Court of Justice—for the regulation of all conflicts between nation-states. In short, the absence of social institutions essential to a global society acts as a barrier to the development and application of scientific knowledge to global problems.

THE TRAGEDY OF THE COMMONS

Given this barrier to the development of a global society, world problems tend ineluctably to take on the proportion of a "tragedy of the commons" so well articulated by the biologist Hardin (1968):

> The tragedy of the commons develops in this way. Picture a pasture open to all. It is to be expected that each herdsman will try to keep as many cattle as possible on the commons. Such an arrangement may work reasonably satisfactorily for centuries because tribal wars, poaching, and disease keep the numbers of both man and beast well below the carrying capacity of the land. Finally, however, comes the day of reckoning, that is, the day when the long-desired goal of social stability becomes a reality. At this point, the inherent logic of the commons remorselessly generates tragedy.
>
> As a rational being, each herdsman seeks to maximize his gain. Explicitly or implicity, more or less consciously, he asks, "What is the utility *to me* of adding one more animal to my herd?" This utility has one negative and one positive component.
>
> (1) The positive component is a function of the increment of one animal. Since the herdsman receives all the proceeds from the sale of the additional animal, the positive utility is nearly +1.
>
> (2) The negative component is a function of the overgrazing created by one more animal. Since, however, the effects of overgrazing are shared by all the herdsmen, the negative utility for any particular decision-making herdsman is only a fraction of −1.

> Adding together the component partial utilities, the rational herdsman concludes that the only sensible course for him to pursue is to add another animal to his herd. And another; and another. . . . But this is the conclusion reached by each and every rational herdsman sharing a commons. Therein is the tragedy. Each man is locked into a system that compels him to increase his herd without limit—in a world that is limited. Ruin is the destination toward which all men rush, each pursuing his own best interest in a society that believes in the freedom of the commons. Freedom in a commons brings ruin to all.

The planet Earth may be viewed as a commons to which Hardin's model is applicable (King, 1979). The herdsman's freedom to pursue his self-interest in the commons, which in the end proves to be self-destructive, is indeed an apt metaphor of the freedom of nation-states to exercise their sovereign powers over resources in the international system. In seeking to maximize its own welfare with little heed for first order—let alone second or higher order—global consequences, each nation-state, like the greedy herdsman, contributes to such problems as the arms race, the depletion of natural resources, overpopulation, and recurrent famines. To make matters worse, even if nation-states were willing to change their grazing habits in the commons, so to speak, there are few institutionalized mechanisms within the scientific community that can provide knowledge-based guidance from the vantage point of the welfare of the commons, i.e., from the perspective of humanity. Instead, there are a multitude of pressures to coopt the scientific and technological community and to channel its fund of knowledge to exclusive *national* interests rather than inclusive *human* interests. Thus, paradoxically, the scientific and technological community, notwithstanding its international ethos, wittingly and unwittingly contributes to the development of the tragedy of the commons.

THE SCIENTIFIC COMMUNITY AND THE EMERGENCE OF A GLOBAL SOCIETY

Unlike nation-states, the scientific community has no territorial boundaries. Its ethos of advancing scientific knowledge elicits the commitment of scientists regardless of national, ethnic, religious, and racial origins. However, because it is a *cultural community* rather than a *territorially-based community,* it has very limited autonomy or resources at its command.

As knowledge grows in a field of science, those participating in the knowledge creation process become members of an ever widening invisible community. In time, this invisible community (including "invisible colleges") becomes more and more visible and formal in character, first within the restricted area of a nation-state and then across national boundaries (Crane, 1972). Thus, scientific and learned societies are formed to facilitate communication and cooperation among members of the scientific community.

Although learned societies and academies in the arts, sciences, and humanities have a venerable history dating back to the schools of Athens modeled after Plato's Academy in the fourth century B.C., it was not until the sixteenth and seventeenth centuries that they greatly increased in number and variety. With the advent of modern science, the Royal Society of London–inspired, as some scholars claim, by Bacon's philosophy of experimental inquiry (Fowler, 1881: 196-197)–was founded in 1662. Shortly thereafter, The Acádemie des Sciénces was established in 1666, and in the early decades of the eighteenth century, the Berlin Academy of Sciences, the Academy of St. Petersburg, and the American Philosophical Society in Philadelphia were organized.

With the exception of the Royal Society, all of these societies tended to pursue their scholarly activities within the confines of their own countries. In an address commemorating the three hundredth anniversary of Isaac Newton, the Russian physicist, Kapitza, refers to Newton's active role in the affairs of the Royal Society, observing that the Royal Society "in contrast to all academies, has still the character of a non-state society. . . . Since the very beginning of its existence, the Royal Society has maintained contact with many foreign scientists. . . . Since Newton's time, the Royal Society has maintained a lively contact with scientific societies the world over" (Ter Haar, 1967: 139). It was not until the nineteenth century, however, that *international* associations, bringing together learned societies and academies as well as individual scientists and scholars from various countries, began to arise. Their growth continued in the twentieth century but was sharply curtailed by the two World Wars. After World War II, there was a marked growth in the international scientific community. UNESCO was instrumental in facilitating the establishment of many international scientific associations, especially in the social sciences. The multitude of scientific associations has since been linked together in two federations: The International Council of Scientific Unions (ICSU) and the International Social Science Council (ISSC). In the various professions which apply science and technology such as medicine, law, engineering and architecture, there are numerous international associations (Evan, 1974, 1975), as well as federations of such organizations, e.g., World Medical Association, International Dental Federation, International Union of Architects, and the Union of International Engineering Organizations.

What impact, direct or indirect, do such international scientific and professional associations (ISPAs) have on world affairs? In a world in which the nation-state is still the dominant political unit, it may seem an exercise in idle curiosity to consider the role of ISPAs in international affairs. However, since World War II, the international system has undergone significant changes such that the "state-centric paradigm" (Nye and Keohane, 1972) is no longer deemed adequate for understanding world events. New transgovernmental and transnational "actors" have appeared in the international arena which may, in due course, transform relations among nations (Evan, 1962; 1976; Burton et al., 1974; Inkeles, 1975). With the establishment of the U.N., over 200 specialized agencies, such as international governmental organizations (IGOs), have emerged. In the intervening decades, three additional significant developments have occurred in the international system: the proliferation of international nongovernmental organizations (Skjelsbaek, 1972); the marked growth of multinational enterprises; and the emergence of various regional organizations, notably the European Economic Community. The traditional state-centric paradigm not only assumed a unitary position on the part of the nation-state but also attributed significance only to interactions among nation-states, thus overlooking the pattern of multiple interactions between nation-states and transgovernmental as well as transnational entities.

ISPAs AS TRANSNATIONAL ACTORS

Why single out ISPAs, only one category of nongovernmental organizations (NGOs) or international nongovernmental organizations (INGOs), for special attention? First, they constitute approximately one-half of all INGOs. Second, in the course of developing and applying scientific and professional knowledge, ISPAs engage in a variety of interactions with other ISPAs, with their national affiliates, with IGOs, with nation-states, regional organizations, and with multinational enterprises. Third, the visibility of ISPAs, particularly of physicists, biologists, economists, and so forth, has been increasing in recent years and is likely to increase in the future. Fourth, in pursuing their activities across national boundaries, ISPAs contribute to the development of a growing body of scientific knowledge and technology. In turn, the growth of scientific knowledge reinforces the ethos of science and strengthens the commitment of scientists to such cultural values as universalism, organized skepticism, disinterestedness, and communality (Merton, 1957). Additionally, advances in technology, to which ISPAs also contribute, affect the level of economic development, military capability, and relative political power of nation-states. In short, ISPAs directly and indirectly affect four dimensions of the international system: the economic, the political, the social, and the cultural.

That nation-states—be they superpowers or mini-states—are not economically independent is a truism. The variety of economic transactions—trade flows, capital flows, labor flows, and technology transfers—that occur between and among nation-states, are by no means random processes. Nor are the patterns of political alliances and coalitions among nation-states a random occurrence. Rather, they are a consequence of defined and articulated interests, national or subnational, and they, in turn, affect policy decisions concerning the exercise of military power among nations and recourse to legitimate authority as expressed in international law.

The social dimension of the international system, though less tangible than the economic and political dimensions, is nonetheless significant. It consists of patterns of recurrent interactions among rank-and-file citizens as well as elites from different countries, as a function of economic transactions, tourism, migration, membership, and participation in intergovernmental and international associations, mass media of communication, and so forth.

The cultural dimension of the international system, while even less visible than the social dimensions, is of vital importance. It refers to the development and dissemination of a set of values, beliefs, and loyalties that can potentially serve either to integrate or to fragment the various actors comprising the international system.

Do ISPAs, in the course of pursuing their several activities, contribute directly or indirectly to international policy decisions of an economic, political, social, and cultural nature? A conference on ISPAs held at the University of Pennsylvania sought to address this question. It had a twofold objective: the identification of the overall functions of ISPAs, and the delineation of their impact on international decision-making.

In the opening chapters by Crane and Kriesberg, ISPAs are depicted as a species of transnational actors. Unlike international governmental organizations (IGOs), they are nongovernmental in conception and in modus operandi. Crane considers three alternative functions performed by ISPAs, emphasizing their network of interactions with a population of IGOs. Kriesberg examines the structural attributes of ISPAS, particularly their age, their national membership and staff structure, and their resources as they affect their potential influence on the decisions of United Nations agencies with which they have a "consultative status." Both Crane and Kriesberg underscore the constraining effect of ISPAs' limited resources on their effectiveness.

That the influence of ISPAs is often quite indirect is reflected in Johnson's essay on the efforts of networks of economists to reform the international monetary system. A more direct effort to apply knowledge to the problems of the world economy is represented in the essay by Klein. Together with econometricians in 13 countries, 4 developing regions, and the socialist countries affiliated with the Council for Economic Mutual Assistance, Klein has established an ongoing research program known as Project LINK. This project seeks to integrate the economic models of developed and less developed countries so as to improve the accuracy of economic forecasting and thus generate policy proposals to enhance the functioning of the world economy. Applying Project LINK, Klein has, for example, developed economic projections for use in decisions by the United Nations Conference on Trade and Development. Although not formally an ISPA, Project LINK, like Johnsons' "networks of economists," is an example of an "international invisible college."

Utilizing a systems analytic model which differs from that of Project LINK, Mesarovic and Hughes develop a "multilevel regionalized world model" in order to represent the world system as a set of interconnected subsystems. In so doing, they are building upon the Club of Rome project, *Mankind at the Turning Point* (Mesarovic and Pestel, 1974). The underlying objective of this model is to train economic planners and policy makers to make decisions within the framework of a regional as well as a global economy.

Unlike the three essays by Johnson, Klein, and Mesarovic and Hughes, which indirectly address the question of the economic impact of ISPAs, the essays by Miles, Feraru, Pentz and Slovo, and King inquire into their political influence. Miles analyzes the political impact of two ISPAs affiliated with the International Council of Scientific Unions (ICSU): the Committee on Space Research (COSPAR) and the Scientific Committee on Oceanic Research (SCOR). Since 1958, COSPAR has investigated the potentially harmful effects of experiments in space and has provided technical advice to the United Nations General Assembly and to the various specialized agencies of the UN. For over a decade, SCOR has had a major impact on intergovernmental decisions concerning the launching of programs in oceanography. However, as regards the United Nations Conference on Law of the Sea, despite its efforts, SCOR has been unable to exert any appreciable influence in protecting freedom of research within a 200-mile coastal limit, let alone in resolving the dispute between the Group of 77 and all other countries. In a similar vein, Feraru compares the effects of various ISPAs on the policy decisions of two other United Nations conferences, one dealing with problems of the environment and the other with problems of human settlements. Feraru suggests that because of the great diversity of interests and resources, ISPAs have a limited capacity to act collectively. However, those ISPAs with relevant expert knowledge are more likely to exert an influence than those lacking in such knowledge.

The "Pugwash Conference on Science and World Affairs," conceived as a nongovernmental forum for technical, nonideological deliberations among scientists from East and West, has accumulated a highly visible and impressive record of policy influence. Pentz and Slovo examine the range of global problems addressed by Pugwash participants and their changing perceptions regarding the mission of this organization. Meeting periodically since 1957, members of Pugwash have developed proposals that have contributed to the drafting of several significant treaties, notably the Test Ban Treaty of 1962, the Nuclear Non-Proliferation Treaty of 1968, and the Convention Barring the Development, Manufacture and Stockpiling of Biological Weapons of 1972 (Pugwash Conferences on Science and World Affairs, 1978).

The Club of Rome is another example of an ISPA that has inspired a great deal of thought about the global consequences of the policy decisions of nation-states. In his essay tracing the origins of the Club of Rome, Alexander King, co-founder, reviews the Club's well-known study, *Limits of Growth* (Meadows et al., 1972), which provoked widespread discussion of the global effects of policies of unrestrained growth for the depletion of nonrenewable energy resources, the degradation of the environment, overpopulation, and the underproduction of food. Of particular significance are the periodic efforts of the Club of Rome, comprising approximately 100 outstanding scientists and industrialists, to engage prime ministers and presidents of various countries in policy discussions of questions arising from its ongoing research.

The three essays by Angell, Carr, and Scoble and Wiseberg seek to trace the social and cultural impact of ISPAs. After distinguishing between social and cultural integration, Angell outlines the histories and impacts of two ISPAs, the International Statistical Institute and the International Council of Nurses. In the process, he considers the question of alternative channels through which ISPAs seek to exert influence. Carr reviews the efforts of teachers' organizations to advance their proposal for the creation of UNESCO at the founding conference of the UN in 1945. Another proposal, promoted by a number of international organizations of teachers to establish an international university, was finally approved by the United Nations General Assembly in December 1973.

In the closing essay of the book, Scoble and Wiseberg assess the role of international nongovernmental human rights organizations such as the International Commission of Jurists, Amnesty International, and International League for Human Rights, in promoting the values of human rights throughout the world. The fact that one of the principal violators of human rights, as Scoble and Wiesberg indicate, are national governments—some of whom may even proclaim their commitment to such values—points up a serious dilemma: How can the values articulated in the Universal Declaration of Human Rights be effectively implemented? In an effort to solve this dilemma, Scoble and Wiseberg document the noteworthy contributions of those ISPAs concerned with the protection of human rights in all countries.

A MODEL OF THE CHANNELS OF INFLUENCE OF ISPAs ON THE INTERNATIONAL SYSTEM

The challenge facing ISPAs is to discover how to enlarge their scope of influence on international policy-making in the future. In other words, how might ISPAs more effectively transform their fund of scientific and technological knowledge into power? A possible answer to this question may be derived from a process of mapping the channels of influence currently being used by ISPAs. With this view in mind, I present in Figure 1 a descriptive model of some of the actual and potential channels of influence.

Out of the continuously growing fund of scientific and technological knowledge come policy proposals by ISPAs and two other entities: national affiliates of ISPAs (or NSPAs, namely, national scientific and professional associations), and international invisible colleges (IICs). NSPAs can serve as an important bridge between ISPAs and national government agencies. Likewise, IICs, which consist of prestigious scientists, technologists and other professionals who are members of informal international networks, can serve as valuable linkages between ISPAs and national as well as international actors.

An examination of the model in Figure 1, in which I have summarized the principal network interactions of ISPAs, leads to the following conclusions: (1) By far the most frequent channel of influence employed by ISPAs is the IGO; (2) the next most frequent is the national government agency; and (3) the other potential channels of influence, however indirect, are grossly underutilized. Of considerable potential significance is the linkage between ISPAs and their national affiliates, on the one hand, and IICs, on the other. Another interorganizational linkage not represented in the model in Figure 1 but of potential value, as Feraru, Crane, Kriesberg, and other contributors make clear, is the need for a higher order of coalition formation among ISPAs if they are to augment their influence on global policy decisions.

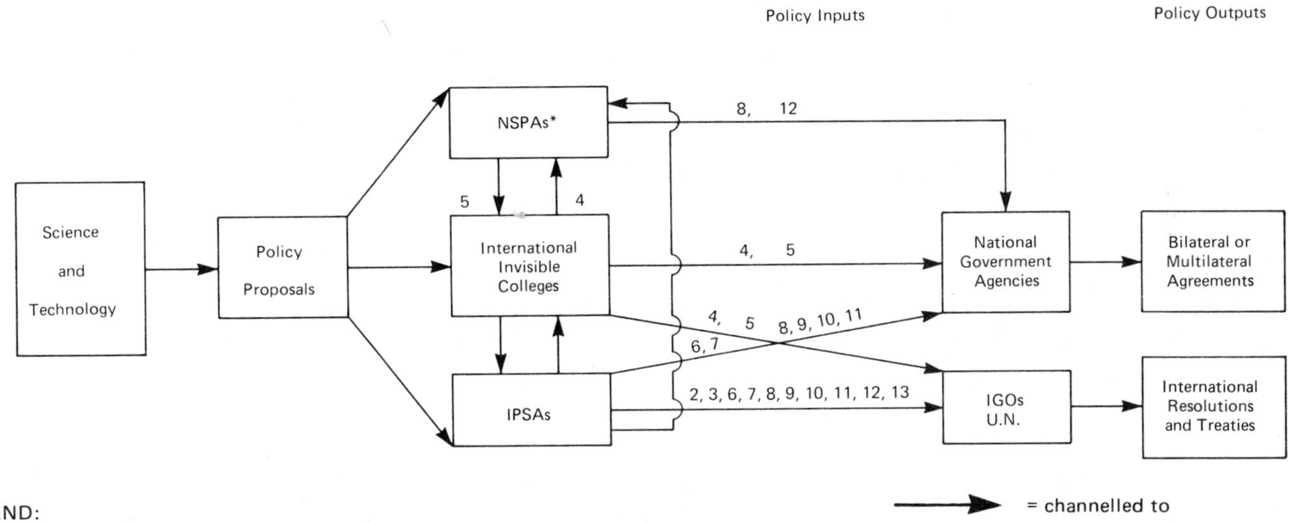

LEGEND:

2 = Ch. 2 (Crane)
3 = Ch. 3 (Kriesberg)
4 = Ch. 4 (Johnson)
5 = Ch. 5 (Klein)
6 = Ch. 6 (Mesarovic and Hughes)
7 = Ch. 7 (Miles)
8 = Ch. 8 (Feraru)
9 = Ch. 9 (Pentz and Slovo)
10 = Ch. 10 (King)
11 = Ch. 11 (Angell)
12 = Ch. 12 (Carr)
13 = Ch. 13 (Scoble and Wiseberg)

Figure 1: A Model of the Influence of ISPAs, NSPAs[a] and Invisible Colleges on International Policy-Making

a. National scientific and professional associations.

TRANSNATIONALISM IN THE SCIENTIFIC COMMUNITY

The barriers to increasing the influence of ISPAs on international policy-making are indeed formidable. Perhaps the most visible barrier is the economic and political dependence of scientists and technologists on their governments and employers. The mere threat of dismissal from a job or termination of a research grant can gravely affect the capacity of scientists and technologists to advocate transnational policies held in disfavor by officials of their own nation-states. The absence of effective international legal restraints on nation-states in penalizing ideological dissidents or "whistle-blowers" severely limits the capacity of scientists, engineers and other professionals to espouse independent and internationalist views on policy issues. The incentives, economic as well as political, are so disproportionately weighted towards conformity that it is scarcely surprising that scientific elites and rank-and-file scientists alike of various countries almost invariably support the nationalist policies of their own governments. If ISPAs were to confront the full implications of this problem, they would endeavor to safeguard the rights of scientists and technologists to freedom of expression, freedom of association, and freedom of research, as provided in the Universal Declaration of Human Rights.[2]

A corollary of the problem of economic and political dependence is reflected in ISPAs' paltry resources to conduct independent global research and to disseminate the findings of such research to policy makers and the general public. The scant resources that are presently available are derived primarily from United Nations agencies such as UNESCO and from the national affiliates of ISPAs. New sources of funds are essential, whether from patent, trademark and copyright royalties, or endowments, for financing transnational policy-related studies on global problems and for making their results known.

Another obstacle to increasing the influence of ISPAs is their tendency to concentrate lobbying efforts at the level of intergovernmental agencies. Although two umbrella-like super-ISPAs are in existence, namely the International Council of Scientific Unions and International Social Science Council, they tend to confine their efforts at policy persuasion to meetings of intergovernment agencies such as UNESCO or World Meteorological Organization. They have yet to develop supplementary strategies of devising channels of communication and interorganizational networks to transmit the policy implications of global studies to their affiliated national member organizations and, through them, to all national governments.

These barriers to increasing ISPA influence will probably remain well-nigh insurmountable in the future unless a concerted effort is made (a) to develop more effective interorganizational influence networks, (b) to acquire independent resources for global research and for policy lobbying, and (c) to institutionalize effective legal safeguards, under the articles of the Universal Declaration of Human Rights, for scientists, engineers, and other professionals, as well as for their national and international organizations, to advocate research-based policies on global issues that are at variance with the national interests of their governments.

If ISPAs were to embark on such a program, they would contribute to the growth of a transnational consciousness within the scientific community, notwithstanding the obstacles to such a development (Schroeder-Gudehus, 1977). Another impetus for the growth of transnational loyalties is the increasingly ominous threat of nuclear war as more nations join the "nuclear club." For this reason, we would do well to keep in mind the prescient global humanist appeal of Russell and Einstein in their manifesto for peace addressed in 1955 to the participants of the first Pugwash (1978: 10-12) conference:

> We are speaking . . . not as members of this or that nation, continent, or creed, but as human beings, members of the species Man, whose continued existence is in doubt. . . .
>
> We have to learn to think in a new way. We have to learn to ask ourselves, not what steps can be taken to give military victory to whatever group we prefer, for there no longer are such steps; the question we have to ask ourselves is: what steps can be taken to prevent a military contest of which the issue must be disastrous to all parties?
>
> There lies before us, if we choose, continual progress in happiness, knowledge, and wisdom. Shall we, instead, choose death, because we cannot forget our quarrels? We appeal, as human beings, to human beings: Remember your humanity, and forget the rest. If you can do so, the way lies open to a new Paradise; if you cannot, there lies before you the risk of universal death.

In the remaining decades of this century, if not longer, humanity will be confronted with the fateful challenge of making the transition from nationalist values to transnational humanist values. If any segment of a society can possibly contribute to the process of transcending the "idols of the tribe," it is the community of scientists, engineers, and other professionals whose work brings them into recurrent contact with colleagues the world over. As the organized expression of this community, ISPAs can potentially make an invaluable contribution to this historical transition. To the extent that ISPAs will succeed in raising the level of transnational consciousness and in persuading national and international actors to apply scientific knowledge in grappling with global problems, to that extent

Francis Bacon's aphorism may yet signal the advent of a new stage in the history of the human species.

NOTES

1. "Realizing the four values [peace, economic well-being, justice, and ecological balance] of global humanism will indeed be difficult. But failing to realize them will be disastrous for millions of coinhabitants of our planet, if not eventually for all human civilization" (Robert C. Johansen, 1978: 407).

2. A promising development along these lines is ICSU's commitment to the Principle of Universality and to the Principle of Free Circulation of Scientists (Brown, 1976: 1). While the former principle affirms the right of membership in ICSU of all scientific organizations in all nation-states, the latter upholds the rights of scientists to leave or enter a country to exchange scientific information or to engage in collaborative scientific endeavors. ICSU's standing Committee on Free Circulation of Scientists is charged with assisting scientists in obtaining visas to attend scientific meetings by asking national scientific organizations to intercede with their governments on behalf of the individuals involved (Walsh, 1976: 589). Brown (1976), president of ICSU, deplores the frequent violation of the Principle of Free Circulation of Scientists. He acknowledges that "the record abounds with refusals of permission to enter countries and refusals of permission to leave countries. The record, unfortunately, also includes refusals of governments, presumably on the basis of political considerations, to permit their scientific communities to join the ICSU family."

REFERENCES

BROWN, H. (1976) "ICSU and humanity," pp. 1-3 in International Council of Scientific Unions, Organization and Activities. Paris: ICSU Secretariat.

BURTON, J. W. et al. (1974) The Study of World Society. Pittsburgh: University of Pittsburgh International Studies Asscciation.

CRANE, D. (1972) Invisible Colleges. Chicago: Univ. of Chicago Press.

EVAN, W. M. (1976) Organization Theory. New York: John Wiley.

--- (1975) "ISA and the internationalization of sociology." Int. Social Sci. J. 27: 385-392.

--- (1974) "Multinational corporations and international professional associations." Human Relations 27, 6: 587-625.

--- (1962) "Transnational forums for peace," pp. 393-409 in Q. Wright et al. (eds.) Preventing World War III. New York: Simon and Schuster.

FOWLER, T. (1881) Bacon. New York: G. P. Putnam.

HARDIN, G. (1968) "The tragedy of the commons." Science 162 (December 13): 1243-1248.

INKELES, A. (1975) "The emerging social structure of the world." World Politics 27 (July): 467-495.

JOHANSEN, R. C. (1980) The National Interest and the Human Interest. Princeton, NJ: Princeton Univ. Press.

KING, A. (1979) The State of the Planet. Elmsford, NY: Pergamon.

LAKOFF, S. A. (1977) "Scientists, technologists and political power," pp. 353-391 in I. Spiegel-Roesing and D. de Solla Price (eds.) Science, Technology and Society. Beverly Hills: Sage.

MEADOWS, D. H. et al. (1972) The Limits of Growth. New York: Universe Books.

MERTON, R. K. (1957) Social Theory and Social Structure. New York: Macmillan.

MESAROVIC, M. D. and E. PESTEL (1974) Mankind at the Turning Point. New York: E. P. Dutton.

NYE, J. S., Jr. and R. O. KEOHANE (1972) "Transnational relations and world politics: a conclusion," pp. 371-398 in R. O. Keohane and J. S. Nye, Jr. (eds.) Transnational Relations and World Politics. Cambridge, MA: Harvard Univ. Press.

Pugwash Conferences on Science and World Affairs (1978) The Pugwash Movement at Twenty-One. Basingstoke, Hampshire: Taylor and Francis (Printers) Ltd.

SCHROEDER-GUDEHUS, B. (1977) "Science, technology and foreign policy," pp. 473-506 in I. Speigel-Roesing and D. de Solla Price (eds.) Science, Technology and Society. Beverly Hills: Sage.

SKJELSBAEK, K. (1972) "The growth of international nongovernmental organizations in the twentieth century," pp. 70-92 in R. O. Keohane and J. S. Nye, Jr. (eds.) Transnational Relations and World Politics. Cambridge, MA: Harvard Univ. Press.

TER HAAR, D. (1967) Collected Papers of P. L. Kapitza, Vol. III. Elmsford, NY: Pergamon.

WALSH, J. (1976) "ICSU: seeking to separate international science and politics." Science 194 (November 5): 587-589.

PART I:

INTERNATIONAL SCIENTIFIC AND PROFESSIONAL ASSOCIATIONS (ISPAs) AS TRANSNATIONAL ACTORS

2

ALTERNATIVE MODELS OF ISPAs

Diana Crane

The number of international scientific and professional associations (ISPAs) has increased rapidly during the twentieth century and particularly during the last two decades, but there have been surprisingly few studies of the phenomenon. In general, the few studies that exist have used the following approaches to understand the activities of ISPAs: (1) the *integration* approach—that the activities of these organizations increase the level of integration between nation-states; (2) the *interest-group* approach—that they exert political influence in the international system; and (3) the *self-interest* approach—that they are primarily concerned with furthering the professional interests of their members. In this chapter, I will review the existing literature on ISPAs in order to show the state of existing knowledge on this subject.

In the concluding sections of my chapter, I will suggest that an appropriate strategy for understanding these three functions of ISPAs is network analysis, since it is apparent that ISPAs participate in many intersecting networks. On the one hand, ISPAs participate in networks of international nongovernmental organizations (NGOs), international governmental organizations (IGOs), national nongovernmental organizations (NNGOs) and national governments. How autonomous are they? Are they influenced more by IGOs or by their national counterparts? As we shall see, ISPAs are often the foci for international informal communication networks of individuals who accomplish their own goals and influence international policies through or in spite of these organizations. Are ISPAs of secondary importance compared to these international informal communication networks? First, however, I will present a brief "demography" of these organizations in order to clarify the nature of the phenomenon that is being discussed.

A DEMOGRAPHY OF ISPAs

If, following Evan (1974), one defines ISPAs as international nongovernmental organizations (INGOs) concerned with the following fields: economics and finance, health and medicine, law and administration, professions and employers, science, social science, and technology, ISPAs constituted 41 percent of the active INGOs (2173) in 1972 (Union of International Organizations, 1974).[1] Between 1909 and 1972, these organizations increased by 1134 percent. Between 1952 and 1972, they increased by 220 percent. By comparison, all other INGOs increased by 185 percent during the same period. The average annual growth rate of ISPAs between 1952 and 1972 was 11 percent (8 percent between 1962 and 1972).

As Table 1 shows, different types of ISPAs have been increasing at different rates during the past twenty years. ISPAs have been increasing most rapidly in the area of technology followed by economics and finance, social sciences and humanities, health and medicine, science, professions and employers, and law and administration. However, the largest numbers of ISPAs are found in health and medicine, science, technology and the social sciences and humanities.

Clearly, many of these organizations are regional rather than international. Judging from the titles of the organizations alone, there are considerable differences in the proportions of regional ISPAs in each field. This proportion is highest in health and medicine and lowest in the social sciences and humanities (see Table 1). One would expect that the impact of regional organizations on international integration or as political pressure groups would be substantially different from that of the international organizations. National participation in these organizations also varies considerably. Skjelsbaek (1971) showed in 1966 that over 50 percent of the national members in all INGOs were developed countries (e.g., North America and Western Europe).

It is also helpful to compare ISPAs with other INGOs. Table 2 shows the percentage of ISPAs and other INGOs by category in 1972. Among the ISPAs, health and science are the biggest categories; among other INGOs, commerce and industry predominate.

TABLE 1 Number, Percent, and Growth of ISPAs by Category: 1954-1974

Category:	*Number (Percent) of Organizations:**			*Percentage Regional in 1974:*	*Percentage Increase:*		
	1954	*1964*	*1974*		*1954-64*	*1964-74*	*1954-74*
Health, med.	101 (28)	150 (27)	288 (30)	38	49	92	185
Science	81 (23)	118 (21)	192 (20)	22	46	63	137
Technology	34 (10)	70 (12)	142 (15)	32	106	103	318
Soc. sci., hum.	38 (11)	67 (12)	124 (13)	15	76	85	226
Profl., empl.	56 (16)	78 (14)	117 (12)	25	39	50	109
Law, admini.	31 (9)	45 (8)	57 (6)	23	45	45	84
Economics, finance	14 (4)	33 (6)	50 (5)	34	136	52	257
TOTAL	355 (101)	561 (100)	970 (101)		58	73	173

SOURCE: Adapted from Table 3, *Yearbook of International Organizations*, 1974, page S33 and lists of ISPAs in the same volume.

*Listed in order of relative frequency in 1974.

TABLE 2 Percentage of ISPAs and Other INGOs by Category, 1972

Category:	*ISPAs*	*Category:*	*Other INGOs*
Health, medicine	12	Commerce, industry	12
Science	8	Intl. relations	7
Technology	6	Religion	5
Social sci., hum.	5	Soc. welfare	5
Profl., empl.	5	Education, youth	5
Law, admini.	3	Sport, recreation	5
Econ., finance	2	Agriculture	4
Total %	41	Transportation	4
N	(901)	Arts	4
		Biblio., docu.	3
		Trade unions	3
		Politics	1
		Total %	58
		N	(1272)

SOURCE: Adapted from Table 3, *Yearbook of International Organizations*, 1974, page S33.

THREE APPROACHES TO ISPAs

Review of the literature on ISPAs and INGOs shows that their activities have been examined using three principal approaches. The first is the *integration* approach. Evan (1974) and Smoker (1965) argue that the activities of INGOs are increasing the level of integration between nation-states. As Evan formulates this interpretation, these organizations are believed to contribute to "the process of normative, economic, organizational, and occupational integration of the international system" by increasing the number of linkages between nation-states. In other words, these organizations increase the number of contacts that individuals from different nation-states have with one another and thereby influence their attitudes and behavior toward members of other countries as well as the frequency of economic and other types of exchanges. Such activities represent interests shared by members of nation-states and thus should serve to counteract tendencies toward polarization along national lines.

A second is the *interest group* approach. As Roosevelt (1970: 284) puts it: "Non-governmental organization is the polite term we use for interest group, pressure group or voluntary association." According to this view, INGOs exist to exert political influence in the international system, primarily upon more powerful actors in the system, such as the IGOs. Two questions are of central interest here: What kinds of influence do INGOs exert upon these organizations? Under what conditions do they achieve their goals?

Third, the activities of these organizations can be understood in terms of pursuing their own *self-interest,* for example, the professional interests of their members. How do the characteristics of ISPAs that are primarily self-interested differ from those that attempt to influence IGOs and INGOs in order to accomplish goals that transcend the self-interests of their members?

Are these approaches mutually exclusive? In other words, if ISPAs are found to be working mainly to serve the interests of their members, does this mean that their political influence is limited or that their activities contribute little to the integration of the international system? This seems unlikely. In fact, the activities of many ISPAs probably contribute to two or more of these goals most of the time. However, for purposes of research, it is useful to analyze these three functions separately.

ISPAs AND INTERNATIONAL INTEGRATION

According to the integration approach, the consequences of increased contact between individuals from different countries through IGOs and INGOs are favorable. Such contacts produce positive attitudes toward the objectives of international organizations and thereby contribute to the integration of the entire system. This viewpoint assumes: (1) that the number of international contacts increases as a result of participation in INGO activities, and (2) that the consequences of these contacts are positive.

Both Smoker and Evan provide evidence that the number of international contacts increases among participants in INGO activities. For example, Smoker (1965) scored categories of INGOs in terms of the numbers of nations belonging to the INGOs in that category. Between the end of World War II and 1960, such scores approximately doubled for most of the categories of INGOs. These figures suggest that such INGOs are increasing the level of international contacts between members of nation states. He also looks at relationships between individual countries and shows that some types of cultural activities, particularly science, health, and technology, increase the bonds between Eastern and Western countries, both large and small. Smoker (1965: 645) suggests on the basis of these data that "joint scientific ventures might be treated as economically seriously as defense projects."

Evan (1974) argues that the large number of international congresses held annually by the ISPAs increases the bonds between international organizations and affiliated national organizations (i.e., the national professional associations in the various nation-states). Evan (1974: 611, 614) says that

> by eliciting participation of national professional associations from various nation-states, the international professional associations un-

intentionally create a network of relationships between nation-states . . . These organizations are socializing their members into the values of transnationalism . . . Thus we hypothesize that these emergent loyalties are raising the level of integration of the international system.

Evan admits that testing these hypotheses would "challenge the ingenuity of a team of social scientists." For example, it would presumably require in part the collection of masses of attitudinal data concerning motivation, loyalties, and preferences.

While these authors generally assume that participants in INGO activities react favorably to their experiences, Nau, in a case study of European technical cooperation, suggests that international contact does not necessarily increase willingness to cooperate in the achievement of shared objectives. Instead, Nau (1972: 521) found that, among scientists from different countries who were attempting to develop cooperative regional programs, "each community covets the opportunity to impress the particular make-up of its own research tradition upon the composition and character of the common program." Cooperation eventually ensued as a result of political negotiation and not from increased integration between scientists.

Judge (1973: 142) presents another version of the integration hypothesis, one which does not make predictions about the consequences of international activities for the system as a whole. He suggests that

many nongovernmental organizations are of major importance to society, either as generators of new values more appropriate to new conditions or through their efforts to protect old values. The nongovernmental sector may be thought of as a value-generating and conserving system.

ISPAs AS INTEREST GROUPS

Other writers have seen these organizations as political interest groups and have been concerned with the kinds of influence which they exert and the conditions under which they exert such influence. In order to evaluate the potential for influence of the ISPAs, it is important to consider whom they might attempt to influence and their resources for doing so. Although only fragmentary data are available, INGOs' financial resources appear to be very limited. A report in 1969 (Judge, 1969: 143) found that "the commonest and most serious weakness of INGOs is inadequate financing." Winings found that the average annual income of INGOs was less than half a million dollars in 1970-71 but he argues (1975a: 468) that

although the financial base of many IGOs and NGOs is shaky by most standards, their capacity to attain stated goals is not seriously hampered, and their impact on the global environment is not insignificant. This is particularly true of organizations that deal in the exchange of scientific and technical knowledge and skills.

In other words, he suggests that an INGO's budget does not necessarily reflect all of the resources at its disposal and this may be particularly true of ISPAs.

It seems likely that in terms of political impact, ISPAs fall into three main groupings: (1) ISPAs which have been created to further the professional or specialized interests of their members and have no need to exert political influence to achieve their goals; (2) ISPAs which were organized in order to pursue a particular cause or interest and which may or may not attempt to exert political influence on more powerful organizations such as national governments and IGOs; and (3) ISPAs that have formal consultative relationships with specific IGOs and that exercise at least some political influence as a result.

It appears that ISPAs which have neither special relationships with NNGOs nor formal consultative status with IGOs are unlikely to be influential. For example, most lobbying for social science programs at UNESCO appears to be conducted by NNGOs such as the American Sociological Association. While the International Council of Scientific Unions (ICSU) plays a much more autonomous role in influencing the natural science programs of UNESCO, it also cooperates with national academies which work within the framework of what their national governments will support, since they in turn appoint representatives to the IGO conferences where decisions are made.

The importance of formal consultative status with IGOs is indicated by Feraru's study of the United Nations Conference on the Human Environment in Stockholm in 1972. She found that many of the INGOs who sent representatives to the conference had little impact, because they did not see the necessity of acting like pressure groups and because their objectives were too diffuse. Significantly, INGOs whose representatives had the greatest impact at the conference were those which could provide scientific and technological expertise, presumably ISPAs. Feraru (1974: 52) found that "contacts among technological specialists in INGOs, IGOs, and governments are likely to be close, important and not very visible." Members of several ISPAs served as expert-consultants in the preparation of Conference reports and recommendations (ICSU, IUCNNR, Society for International Development, and ISSC). Feraru concludes that as far as the UN Environmental Program is concerned, the INGOs most likely to influence it were those that had scientific and technological expertise or those that were able to mobilize support for the Program among influential publics in the countries that will provide financial support for the organization.

While INGOs can exercise a policy influence if they can develop special consultative relationships, either formal or informal, with IGOs, they are unlikely to be able to coordinate their interests and work together to present a common program to the IGOs. Todd (1973: 45) indicates why lack of cooperation was the case at the Stockholm Conference:

> Representatives of scientific and professional [I] NGOs tended to see things differently from those of long-standing conservation-oriented [I] NGOs, while the representatives of citizen-action type [I] NGOs tended to see things still another way. Disagreements arising from these differences among the [I] NGOs were compounded by such things as the industrial or non-industrial status of the [I] NGO representatives' home countries and the languages used by the participants in the [I] NGO meetings.

Roosevelt also sees the majority of INGOs as relatively uninfluential and provides additional reasons for this state of affairs. Roosevelt (1970: 285) sees many of them "as dominated by a relatively small percentage of their memberships. . . . This gulf between membership and leadership undoubtedly diminishes the organization's influence." Some leaderships have difficulty maintaining the interest and commitment of their members.

Probably the most influential INGOs are those that have formal consultative relationships with specific IGOs. Feraru (1974: 52) comments that formal consultative status is not necessary for an INGO to have access to an IGO but she says that "most INGOs with an interest in some aspect of the UN or the specialized agencies' programs seem to acquire consultative status eventually."

Examination of the data included in the *Yearbook of International Organizations* (Union of International Associations, 1974) concerning the consultative relationships of INGOs shows that only 11 percent of ISPAs have such relationships to any significant extent. Of other INGOs, 24 percent have such relationships. The 28 ISPAs that have high scores in terms of the importance of their consultative relationships with IGOs[2] are distributed fairly evenly over the seven categories which we have used to define ISPAs. Technology, health, and professional ISPAs are most frequently represented, followed by science, law, and the social sciences. There are no ISPAs from the category of economics and finance in this group.

The relationship between the Scientific Committee on Oceanic Research (SCOR), a committee of ICSU which functions as an independent INGO, and the Intergovernmental Oceanographic Commission (IOC) can serve as an illustration of a formal consultative tie between an ISPA and an IGO. SCOR is the official scientific advisory body for the IOC, and has more than 30 national committees from which it can seek advice. These national committees assure that the organization is kept in constant touch with the latest scientific developments. From time to time, special working groups are established, sometimes in conjunction with advisory committees of other ICSU bodies. Wooster (1973) concludes that "the program of IOC has been significantly influenced by SCOR advice."

Another element that plays a role in technical, scientific, and social scientific fields is that of the invisible college, the informal network of experts who are in contact with one another but who do not share a single organizational affiliation. Such informal groups, rather than the ISPAs, may actually be the most important actors in such situations. Brenner (1975: 774) shows that the IOC was created by an invisible college of oceanographers. He says:

> IOC was the brainchild of a group of Anglo-American scientists who conceived of a global body to promote scientific studies of the oceans and to coordinate cooperative ventures in advancing this goal. . . . It was an opportune time to undertake such institution-building. Oceanography then truly was an intimate scientific club, most of whose major researchers knew each other personally, and who in their academic capacities embodied oceanography. They were in a position to define purposes, and to outline the programmatic instruments for achieving them. . . . With the sympathetic support of their governments and using ICSU as the instrument to marshal international scientific support for their initiative, they succeeded in founding IOC.

He also shows that the IOC is a highly complex and decentralized organization which engages in a great variety of activities. Brenner (1975: 781) suggests that coordination of these activities is provided by the invisible college, who perform many different and overlapping roles as "delegates, coordinators, advisers or consultants" on the various committees and working groups of the organization.

> Prominent oceanographers often serve simultaneously as directors of important national research efforts and members of government-appointed delegations, while furnishing, in one consultative capacity or another, the expert advice on which IOC depends. Moreover it is not uncommon for a UN administrator to have been previously an official of a national science agency or researcher and to return to that role after the expiration of his UN contract.

Even those scientists who become administrators on the IOC's permanent staff benefit in the performance of their roles from their "network of ties with the international community of oceanography." Recent attempts to reorganize the IOC were also instigated by members of this invisible college.

However, while the existence of this invisible college provides a certain amount of coordination in a highly fragmented situation and helps to "minimize incomprehension and mute conflict," it does not mean that these scientists pursue genuinely international or transnational interests. Instead, Brenner argues that the primary concerns of these scientists are their own research and professional interests. In administrative roles, they are preoccupied with the protection of bureaucratic interests and with the development of organizational programs. Only a "handful of statesmen of science" attempt to assess the situation on a transnational basis and to make recommendations for organizational reform. Brenner (1975: 803) says:

> The scientist's intellectual impulse toward organizational order and clarity of purpose is most often limited to that particular sphere of endeavor with which he has professional familiarity, and the program component of the specialized agency with which he deals.

Burt (1971) in his study of the participation of IGOs, INGOs, and NNGOs in UN decisions on birth control found that commissions of experts played important roles in three out of four UN agencies (ESA, ECOSOC, and WHO). For example, he (1971: 128) says of the World Health Organization (WHO) that:

> technical-level ad hoc scientific committees are the sun around which WHO planets, i.e., programs, revolve, in decision orbits. Presumably, once technical experts . . . recommend decision steps, then members of secretariat technical systems tend to concur, and counterpart specialists in national health ministries of donor governments move toward financial and other support.

ISPAs such as the International Union for the Scientific Study of Population Problems (IUSSP) appear to play secondary roles in this system. The latter is seen as sending representatives to meetings of expert committees and sponsoring conferences but actual decisions are taken by the expert committee.

Burt also suggests that an invisible college of experts plays an important although indirect role in coordinating the activities of the expert committees. He (1971: 150-151) asks:

> Do coalitions of specialists spread out across all three types of organizations (UN, government, and private) form a network which exerts a greater impact upon decisions than do either vertical or horizontal dimensions of organizational structures? . . . Can specialists be thought of as functioning at the interstices of formal substructures of organizations, providing a fundamental information channel through which decisions flow?

He notes that American demographers participating in the activities of these agencies refer to themselves as an "establishment." Burt (1971: 54) also wonders:

> Are a relatively small number of persons trained in the speciality, or discipline, of demography—not just in the United States, but across the world—united by professional norms? Does the wearing of the same professional "school tie" (a) prompt them to 'think alike' in technical decisions, and (b) prompt non-specialists found in decision networks, therefore to defer to demographers' "collective wisdom?"

In the area of international fisheries' management, Johnson (1975) also found that ISPAs played minor roles. In this situation, technocrats, usually with scientific training, make national and international policy on national and international fisheries' commissions. They generally have more influence than experts who are not in such positions, although the latter do exert considerable influence on international scientific committees serving international fishery commissions. The conjunction of national commercial interests with the need for international regulation may explain these patterns in part.

These studies show that it is not necessarily an ISPA which exerts political influence but the expert committee which may or may not be affiliated with an ISPA. The invisible college which cuts across all the organizations involved, both IGO and INGO, plays an important role in integrating the fragmented IGO programs. These studies also suggest that the political interest group hypothesis is insufficient for comprehending the complexities of interorganizational relationships in this sphere.

ISPAs AS SELF-INTERESTED ACTORS

ISPAs can also be seen as primarily pursuing the interests of their members rather than transnational goals, although the latter may be accomplished indirectly as Evan's (1975) study of the International Sociological Association shows. Kriesberg (1972) states that

> most INGOs direct their activities for their own members: facilitating the day-to-day conduct of the members, for example, by providing and exchanging information. Members may also be served by an organization which represents their interests vis-à-vis other groups, e.g., an adversary relationship.

He found that 96 percent of INGOs direct their activities to serve their own members. Eighteen percent did so by representing the interests of members vis-à-vis others.

Unfortunately, our information about the great majority of INGOs that direct their activities toward the interests of their members is very slight.

Roosevelt (1970) says that INGOs are formed "for pragmatic reasons, social pressure or idealism" but he does not attempt to estimate the proportions in each of these categories.

Some ISPAs which are concerned with their members' professional or specialized interests exclusively can only survive by mobilizing support from other organizations, usually NNGOs or IGOs. For example, let us examine the case of three international social science organizations: the International Sociological Association, the International Union for the Scientific Study of Population, and the International Social Science Council. Shortly after World War II, these organizations were created by UNESCO, which has continued to finance them ever since (Glick, 1972). These ISPAs were threatened with loss of financial support and imminent demise when the American government took the position in 1964 that UNESCO's social science budget should not be used in this way. It was the American Sociological Association that came to the rescue, sending representatives to the U.S. Department of State to argue in favor of these social science ISPAs. Eventually, the American government reversed its position. Thus we have ISPAs obtaining funds for maintenance from an IGO by means of a National Scientific and Professional Association (NSPA) and a national government.

One can only speculate concerning the characteristics that differentiate self-interested ISPAs from those that attempt to perform broader roles in the international community. One might hypothesize, for example, that, in general, ISPAs which are created by IGOs will have broader interests than those which are created by NSPAs. International federations of national federations will presumably have broader interests than ISPAs which represent national societies, and those in turn will have broader interests than those which represent individual members only. Significantly, only three ISPAs which represent individual members rather than national societies or federations play important roles as consultants to IGOs.[3]

NETWORK ANALYSIS AS A STRATEGIC METHODOLOGY

None of the approaches which we have reviewed explain ISPA activities entirely satisfactorily. To date, most studies of ISPAs have been case studies of individual organizations. As such, they provide a wealth of historical detail but little systematic or comparable information about how these organizations interact with other organizations. It is clear that there is not, as yet, an active research tradition devoted to the study of ISPAs. In other words, the models that we have discussed have not attracted the interest of a substantial group of researchers who interact with one another in the production of new knowledge in this field. One might speculate that this has not occurred because none of these models adequately comprehends the complexity of the relationships among ISPAs and between ISPAs and IGOs.

The complexity of these relationships can be grasped by examining the enormously intricate consultative relationships between even a small subsample of these organizations. Networks of consultative relationships between IGOs and ISPAs can be examined to find out what they suggest about the relationships between these organizations.[4] I will consider only high-level consultantships[5] to a group of 13 IGOs, and include only ISPAs that have two or more consultantships of this sort. Regional organizations will be excluded entirely, since these organizations probably constitute a separate and distinct social system as far as political influence is concerned.

This analysis does *not* assume that the existence of a consultative relationship between an ISPA and an IGO implies that the former exerts an influence upon the latter. On the contrary, it attempts to clarify the conditions under which these consultative relationships are operating and thereby to obtain some indication of the likelihood that influence will be exercised. These complex networks are further complicated by the activities of other types of organizations, such as INGOs without formal consultative relationships, NGOs, and particularly national governments.

Out of 970 ISPAs, only 49 have high-level consultantships with two or more IGOs and only 9 have such consultantships with 4 or more IGOs. Most ISPAs with 2 to 4 consultantships of this sort (36 out of 40) are linked to one or more of the big three IGOs, ECOSOC, UNESCO, and WHO, which have the largest number of major consultantships (Union of International Associations, 1974: Table 2). Given the density of the networks involving the "big three," one can infer that most ISPAs exert little influence upon these IGOs.

It seems plausible that an ISPA would be most likely to influence IGOs other than the "big three." However, it is primarily the ISPAs that have more than four consultantships such as IOS, ICSU, and IUCNNR, that have ties with these organizations.

It is also useful to compare ISPAs with other INGOs. Network analysis can provide some insight into the system formed by links between these other INGOs and the IGOs. The non-ISPA system is even more heavily concentrated around a few IGOs, in this case, ECOSOC, UNESCO, and UNICEF (see Table 3). Out of 66 INGOs with two or three consultantships, only one was not linked to at least one of these organizations and only 27 percent had at least one consultantship with one of the other IGOs. This suggests that the activities of this group of INGOs are also

TABLE 3 Number of Consultantships with IGOs: With ISPAs and Other INGOs

Number of Consultantships Per IGO:	*ISPA Consultantships*	*Other INGO Consultantships*
WHO	27	15
ECOSOC	24	84
UNESCO	21	63
UNICEF	14	61
WMO	10	4
ITU	9	7
UNIDO	7	10
FAO	6	23
IAEA	6	8
ICAO	5	8
UNCTAD	4	12
IMCO	3	8
ILO	1	5
Number of Organizations	49	91

SOURCE: Based on Table 2, *Yearbook of International Organizations*, 1974: S10-S24.

directed primarily toward these top three IGOs. These data, combined with similar data concerning ISPAs, suggest that the entire INGO-IGO system may be overly centralized.

Does this analysis of the entire network of relationships between ISPAs, INGOs and IGOs help to clarify what we learned from our review of earlier studies which were based primarily on analyses of the interactions among dyads or members of small segments of this network? The density of the network surrounding organizations such as ECOSOC and UNESCO suggests indirectly why many INCOs pursue their own interests, rather than attempting to exert political influence. Their opportunities for influencing these organizations are limited by heavy competition. It would seem that the proposals of Judge (1971) and Speeckaert (1971) for more cooperation among INGOs on a highly flexible basis (in the form of temporary associations and potential associations) are justified. So much activity is centered upon a few IGOs that it is probable that only a few of the most important INGOs have any impact whatsoever.

Previous findings that individuals, experts, and special committees were more able to exert an influence upon IGOs than many ISPAs are not surprising in view of the networks we have described. In other words, given the competition among consultative organizations, it is possible that only those with informal ties who are able to develop "special relationships" with staff of IGOs are likely to have any influence upon IGO secretariats. The predominance of scientific and technical ISPAs among those which have many consultantships may be due to advantages accrued to such organizations by the tendency of their members to develop informal networks of communication and influence.

Examination of networks of consultative relationships between IGOs and ISPAs can provide a kind of "aerial" view of ISPA-IGO relationships. While it cannot reveal the content or importance of these relationships, it suggests which organizations merit further study. For example, a study of the nine ISPAs with four or more major consultantships would be of considerable interest. It is noteworthy that the three organizations with six and eight consultantships apiece are all scientific and technical associations. In the group of nine, scientific, technical and professional organizations predominate.[6] All nine have consultantships with ECOSOC, eight out of nine have consultantships with UNESCO, and six out of mine with WHO. What do they contribute to these IGOs? How much do they influence or coordinate the activities of these IGOs? To what extent do these ISPAs play a role which has been identified in interorganizational studies as mobilizational coordination? Do these ISPAs induce IGOs to undertake certain kinds of activities rather than simply respond to requests for consultation from IGOs?

It seems plausible that if any ISPAs engage in "mobilizational coordination" in the international system, these nine ISPAs would be most likely to do so. Warren (1974: 72) describes "mobilization coordination" as follows:

> In these instances, a single organization sets out to pursue some objective. . . . But it recognizes that it needs the support of other organizations in order to accomplish its objectives. . . . It sets about systematically interacting with other organizations in a decision-making process which, if successful, adds up to the securing of the help it needs from other organizations. . . . It performs, in other words, an entrepreneurial function in the sense that it gathers together the resources and forges the ad hoc relationships that enable it to pursue its objective.

In fact, there is some evidence that ICSU has on various occasions attempted to initiate activities for IGOs. For example, Brenner (1975: 774) states that ICSU played an important role in the founding of an IGO, the IOC. A group of oceanographers who saw the need for such an organization "used ICSU as the instrument to marshal international scientific support for their initiative." Glick (1972), a sociologist, writes enviously of ICSU's role in formulating UNESCO's science programs. Wooster (1973) describes ICSU as "the principal protagonist on the nongovernmental side" in dealing with the problem of marine pollution. Because the problem is interdisciplinary, ICSU has attempted to promote interdisciplinary activities by developing interunion commissions among its many member unions and interdisciplinary special committees. Feraru (1974) in her study of the involvement of INGOs in the 1972 UN Conference on the Human Environment found that, using measures of participation and activity and peer ratings, ICSU was one of the three INGOs most involved in UN environmental activities. Another of the three was IUCNNR, which also belongs to this group of ISPAs with large numbers of consultantships.

What roles do the remaining 40 ISPAs play? It is possible that certain of these ISPAs coordinate the activities of two IGOs in relation to a specific problem area. For example, a few ISPAs, such as International Astronomical Union (IAU) and International Union of Radio Sciences (IURS), concentrate their activities on pairs of smaller IGOs, such as World Meteorological Organization (WMO) and International Civil Aviation Organization (ICAO). Similarly, International Electrotechnical Commission (IEC) is linked to International Telecommunication Union (ITU) and Intergovernmental Maritime Consultative Organization (IMCO) (see Figure 1). Their activities would be especially interesting to examine in depth. ISPAs with at least two links to small IGOs (and a third with one of the big three) such as ICPO, ICID, ISSS, IUPDEE, IUNS, and ICUBE should also be studied (see Glossary of Abbreviations). These organizations are again predominantly scientific and technological (six out of ten); economics and social science ISPAs are not represented. Further research might examine how these organizations differ in their activities and mode of organization from the "mass" of ISPAs which do not have such consultantships.

In other words, a network analysis of the consultative relationships between ISPAs and IGOs can be useful as a way of generating hypotheses concerning their interrelationships and testing them in a preliminary fashion. Social networks are important elements of modern societies because they represent channels for transmission of information, influence, and expertise, and for exchange or allocation of rewards and resources. Absence of social networks or the existence of networks which transmit information, influence, or expertise inefficiently are responsible for many of the malfunctions in large, modern, complex societies. The characteristics of the network itself can affect the extent to which transactions will take place within it, and thus the extent to which the actors in the network will be able to accomplish their goals.

CONCLUSION

Although there is as yet no active research tradition concerning the study of ISPAs, three distinct approaches to the subject can be identified

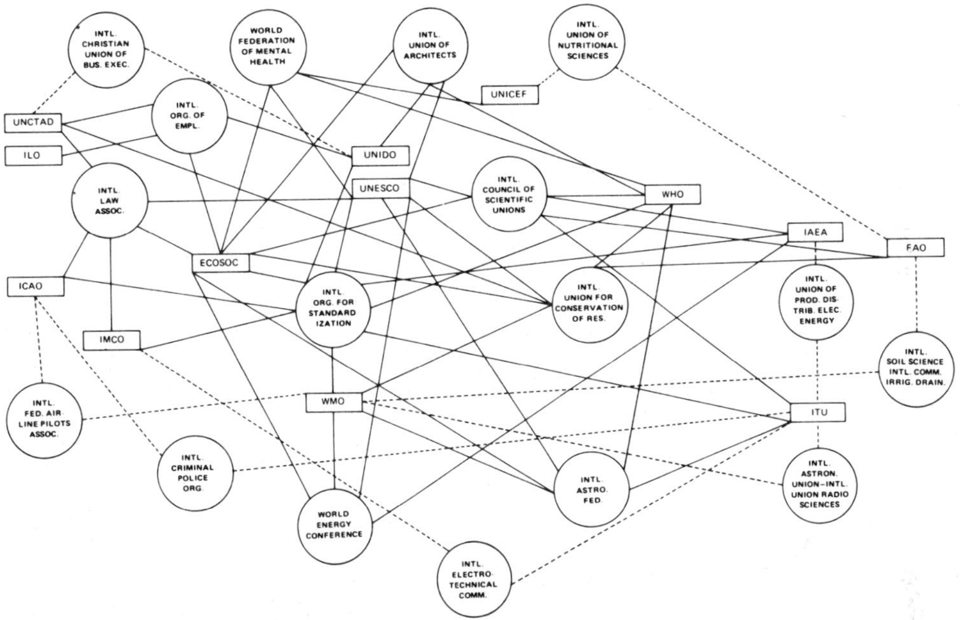

Figure 1: Network of 13 IGOs Created by Consultantships with 9 "Major" ISPAs and 10 ISPAs with Consultantships Primarily with "Minor" IGOs[a]

a. IGOs are shown as squares and ISPAs as circles. The major IGOs (those with the largest number of consultantships) are ECOSOC, UNESCO, and WHO. Major ISPAs are those that have four or more consultantships, according to the **Yearbook of International Organizations**, 15th edition. Ties between 'minor' ISPAs and 'minor' IGOs are shown by dotted lines. All other links are represented by solid lines.

in the literature. These approaches vary in the amount and kind of impact they attribute to ISPAs. ISPAs that are self-interested actors have a minimal impact on international policy-making and on international integration. ISPAs that are concerned with furthering the interests of a particular group may indirectly contribute to international integration. ISPAs that are self-consciously seeking to raise the level of international integration may, indirectly, contribute to international policy-making.

We have suggested that a better understanding of the factors affecting the impact of ISPAs could be obtained if the problem were reconceptualized in terms of social networks. Access to social networks which represent channels for transmission of information, influence, and expertise, and for exchange or allocation of rewards and resources, is an important factor in determining the influence an organization can exert on its environment. Reconceptualization of these approaches in these terms would extend their usefulness and might attract a sufficient number of researchers to the area to create a viable research tradition. Developing methods and models for understanding the complexity of these formal and informal networks offers a challenging set of problems for further research.

NOTES

1. This total does not include EEC/EFTA NGOs, national NGOs with UN consultative status, or short entries. See *Yearbook of International Organizations,* 1974: 533.

2. Based on pages S25-S28, *Yearbook of International Organizations,* 1974. The point system used in the Yearbook is designed to reflect the complexity of different grades of consultative relationships. In the case of ECOSOC, ILO, FAO, and UNESCO, 3 points were given to the principal category of consultantship, 2 to the second, and 1 to the others. For the other agencies, 2 points were given for each relationship. The maximum number of possible points is 34.

3. For example, have scores of 6 or more, using the *Yearbook's* point system.

4. Based on the detailed breakdown of consultative relationships between INGOs and the United Nations specialized agencies in the *Yearbook of International Organizations* (Union of International Associations, 1974: S10-S24).

5. For example, those scored 2 or more in the *Yearbook's* scoring system. See footnote 2.

6. Economics and social science ISPAs are not represented. Among the 40 ISPAs with two or three consultantships, health and medicine ISPAs predominate (14), technology is next with 7, and the remaining ISPAs are divided almost equally among law, science, social science, and professions-employers. Economics and finance are represented only once.

REFERENCES

BRENNER, M. (1975) "The Intergovernmental Oceanographic Commission and the Stockholm Conference: a case of institutional non-adaptation." Int. Organization 29 (Summer): 771-804.

BURT, J. C. (1971) "Participation of government and non-government organizations in UN decisions on birth control: a problem in organizational theory." Ph.D. dissertation, City University of New York.

EVAN, W. M. (1975) "The International Sociological Association and the internationalization of sociology." Int. Social Sci. J. 27: 385-393.

--- (1974) "Multinational corporations and international professional associations." Human Relations 27, 6: 587-625.

FERARU, A. T. (1974) "Transnational political interests and global environment." Int. Organization 28 (Winter): 31-55.

GLICK, C. E. (1972) "ASA-UNESCO relations in retrospect and prospect." Presented at the Meetings of the American Sociological Association, New Orleans, August.

JUDGE, A.J.N. (1973) "Principles of transnational action—an attempt at a set of guidelines." Int. Associations 25: 138-144.

--- (1971) "Wanted: new types of social entity. Part I: the role of the 'potential association'." Int. Associations 23, 3: 148-152.

--- (1969) "Evaluation of international organizations—the need, current efforts, qualitative and quantitative methods." Int. Associations 21, 3: 141-147.

KRIESBERG, L. (1972) "International NGOs and transnational integration." Int. Associations 24, 11: 521-525.

NAU, H. R. (1972) "The practice of interdependence in research and development." Int. Organization 26: 499-526.

ROOSEVELT, C. (1970) "The politics of development: a role for interest and pressure groups." Int. Associations 22, 5: 283-289.

SKJELSBAEK, K. (1971) "The growth of international nongovernmental organizations in the twentieth century." Int. Organization 25 (Summer): 420-442.

SMOKER, P. (1965) "A preliminary empirical study of an international integrative subsystem." Int. Associations 17, 11: 638-646.

SPEECKAERT, G. P. (1971) "Les associations momentanées d'organisations internationales." Int. Associations 23, 4: 205-217.

TODD, J. E. (1973) "The role of NGOs at United Nations Conference on the Human Environment." Int. Associations 25: 43-45.

Union of International Associations (1974) Yearbook of International Organizations, 15th edition. Brussels: Author.

WARREN, R. L. (1974) The Structure of Urban Reform. Lexington, MA: D. C. Heath.

WININGS, T. A. (1975a) "The budget: a dimension of IGO-NGO growth." Int. Associations 27, 10: 468-474.

--- (1975b) "Income and expenditure pattern among IGOs and NGOs: a preliminary search for explanatory variables." Int. Associations 27: 543-546.

WOOSTER, W. S. (1973) "Interactions between intergovernmental and scientific organizations in marine affairs." Int. Organization 27 (Winter): 103-113.

3

VARIETIES OF ISPAs: THEIR FORMS AND FUNCTIONS

Louis Kriesberg

International Non-Governmental Organizations (INGOs) have increased greatly in number and size since World War II. Many observers have noted this and speculated about possible implications (Angell, 1969; Feld, 1972; Evan, 1972; Skjelsbaek, 1972; Kriesberg, 1972). To what extent do INGOs help form a global community of shared understandings or a cross-cutting set of occupational and other interest groups? To what extent do they help deal with global problems transnationally or serve the interests of their members?

In this chapter, I shall attempt to answer these questions by focusing on one set of INGOs: the International Scientific and Professional Associations (ISPAs). My specific objectives are threefold: First, I shall describe the organizational characteristics of ISPAs compared to other INGOs. Second, I shall examine variations among ISPAs in organizational characteristics relevant for their transnational import; and I shall analyze the correlates of those organizational characteristics. Third, I shall reflect upon the possible implications of these and related findings for our understanding of the way ISPAs do and might function in the emerging world society.

AUTHOR'S NOTE: This chapter utilizes data from the 1972-1973 *Yearbook of International Organizations.* I appreciate the support from Syracuse University and the University of Pennsylvania for the coding of the data and William Evan's urging to have it done.

The data for this chapter are largely drawn from information coded from the *Yearbook of International Organizations* (Union of International Associations, 1973). We coded the information from the 1972-1973 edition of the *Yearbook* for purposes of this analysis and to retain comparability with the data previously coded from the 1962-1963 *Yearbook.* This information has been supplemented with personal interviews with officers of several ISPAs.

Before we describe how ISPAs compare to other INGOs, we must set forth the definition of ISPAs used in this chapter. ISPAs are one type of INGO. Like other INGOs, they have members from at least three different countries, and the members are national associations as well as individuals. Evan (1974) has defined ISPAs as INGOs concerned with the following fields: economics and finance, health and medicine, law and administration, professions and employers, science, social science, and technology. This utilizes the typology given in the *Yearbook.* In this chapter, the types given in the *Yearbook* are combined with additional information coded from the *Yearbook* to define ISPAs operationally. Some INGOs in the fields listed above do not involve scientists or professionals working in those areas. On the other hand, professionals and scientists may belong to INGOs in other fields. In coding the *Yearbook* data, therefore, we used information from the title of the organization and from a description of it to try to characterize the INGO as involving people in various occupational roles. For purposes of this analysis, we divided INGOs into three categories: (1) those in which members belong in terms of professional occupational roles; (2) those in which members belong in terms of nonprofessional occupational roles; and (3) those in which people are members in terms of nonoccupational roles.[1] These references were often based upon little information, and in many cases the information was inadequate to make any categorization.

The information on the occupational role of the members was combined with the field or type of INGO to derive a classification of ISPAs. The INGOs were divided into three categories: those which had members who were predominantly in professional roles, those which had members in nonprofessional occupational roles, and those which had members predominantly in nonoccupational roles. The way these two classification systems were combined is presented in Table 1. This provides the basic categories for analytic purposes: ISPAs-INGOs, Non-ISPA-Occupational INGOs, and Nonoccupational INGOs. The distribution of these INGOs by field of activity is presented in Table 2.

As can be seen in Table 2, ISPAs, as defined here, are most likely to be in the Health and Science fields (about 40 percent). Another 10 percent are in the Social Sciences, Humanistic Studies field. By far the largest

TABLE 1 Constructing Categories of ISPA and Non-ISPA INGOs by Combining Occupational Role of Members and INGO Field of Activity

	INGO FIELD			
Occupational Role	*Bibliography, and so on; Social Science; Law; Science; Health*	*Professions; Employees; Trade Unions, Economics; Commerce; Agriculture; Transportation; Technology, Education; Youth; Arts; TV*	*Religion; International Relations; Politics; Social Welfare; Sports*	*Totals*
Professional	459[a]	352[a]	142[a]	953
Non-Professional Occupational	96[a]	465[b]	23[b]	584
Occupation Not Ascertained	129[a]	237[b]	56[b]	422
Nonoccupational	39[c]	90[c]	268[c]	397
No information	62[a]	216[b]	53[c]	331
Totals	791	1,366	549	2,706

(a) ISPA-INGOs
(b) Non-ISPA Occupational INGOs
(c) Nonoccupational INGOs

TABLE 2 INGO Categories by Field of Activity

Field of Activity	*ISPA-INGO*	*INGO CATEGORY* Non-ISPA Occupational	*Nonoccupational*
Bibliography, Documentation, Press	6.0	0.0	.9
Religion, Ethics	4.4	.8	13.3
Social Sciences, Humanistic Studies	9.7	0.0	1.6
International Relations	1.8	2.2	23.1
Politics	.2	.7	4.4
Law, Administration	4.9	0.0	.9
Social Welfare	1.9	2.5	14.4
Professions, Employers Trade Unions	7.6	11.1	2.0
Economics, Finance	1.0	3.3	1.1
Commerce, Industry	4.6	46.2	4.4
Agriculture	3.2	4.6	2.0
Transport, Travel	1.5	6.8	1.3
Technology	5.6	7.4	1.1
Science	16.7	0.0	1.1
Health	22.9	0.0	4.2
Education, Youth	2.0	7.5	6.4
Arts, Literature, Radio Cinema, TV	2.9	5.0	1.6
Sports, Recreation	3.2	1.7	16.0
Totals %	100.0	100.0	100.0
(N)	(1240)	(997)	(450)

proportion of Non-ISPA Occupational INGOs are in the Commerce, Industry field. The Nonoccupational INGOs are more diversified in field: International Relations, Sports, Recreation, Religion, Ethics, and Social Welfare. The concentration of ISPAs in the Health and Science areas does not simply reflect the preponderance of professional workers in those fields. The law and the ministry, for example, have large numbers of persons but few ISPAs. The relative proliferation of INGOs in the Health and Science fields partly reflects the universal language of discourse among workers in the fields, shared ways of working, and also greater possibilities of mutually beneficial professional activities.

ISPAs AND OTHER INGOs

In this first part of the chapter, we examine the characteristics shared by ISPAs, compared with other INGOs. We seek to answer questions such as: Do ISPAs have fewer countries as members than do other INGOs? Is their size and structure different? Is their scale of activities relatively small? Are they more or less likely than other INGOs to have links with additional international organizations?

We begin by considering the aims of the INGOs. Since they are purposive organizations, we expect that their goals affect organizational activities and forms. First, however, what are the stated objectives of INGOs? Most of them (65 percent) mention the objective of facilitating the normal day-to-day activities of members; for example, increasing skills of members and exchanging information. This is most likely to be mentioned by the ISPAs (71 percent) and also likely to be mentioned by the Non-ISPA Occupational INGOs (66 percent); it is least frequently mentioned by Nonoccupational INGOs (46 percent). The other relatively frequently mentioned objective is the promotion of conditions which serve the interests of members (42 percent). ISPAs and Nonoccupational INGOs are most likely to mention it (47 percent) and Non-ISPA Occupational INGOs least likely (33 percent).

On the whole, INGOs have as their primary stated objective helping members to carry on their normal activities. This objective is particularly marked for ISPAs. Other objectives are relatively more often cited by other kinds of INGOs. Thus, Nonoccupational INGOs are most likely to mention trying to change or maintain particular social conditions directly affecting nonmembers (52 percent), but this is not mentioned often by other INGOs (fewer than 20 percent). Non-ISPA Occupational INGOs disproportionally, but not very frequently, cite representing the interests of members against others (27 percent, while fewer than 10 percent of the other INGOs mention it).

References to providing services and assisting nonmembers are even less frequently mentioned. Improvement of conditions of nonhuman beings or the protection of the environment is rarely reported as an objective.

In addition to the stated aims, we coded the activities which the INGOs reported conducting. Consistent with their goals, nearly all ISPAs and most other INGOs reported exchanging information, holding congresses and institutes, and exchanging visits. Most ISPAs and almost as many other INGOs also mentioned engaging in joint efforts, for example, coordinating research. More than one-third of the ISPAs and other INGOs also mentioned providing services for members, for example, abstracting services, libraries, and training programs. More than one-fourth of each INGO category also reported developing common standards or nomenclature. Finally, a small proportion of INGOs reported providing mediators.

TABLE 3 ISPA-INGO Categories by Year of Establishment

Year Established		*ISPA*	*INGO Category* *Non-ISPA Occupational*	*Nonoccupational*
1914 or earlier		8.2	4.9	12.1
1915-1929		10.9	6.0	11.9
1930-1939		6.4	4.0	4.7
1940-1949		12.1	9.9	14.8
1950-1954		15.6	14.6	16.1
1955-1959		16.5	26.5	14.2
1960-1964		17.3	22.0	15.3
1965-1972		13.0	12.0	10.8
Totals	%	100.0	99.9	99.9
	(N)	(1090)	(876)	(379)

Before going on to examine the size and structure of ISPAs, we should note when they were established. This will indicate their age, which is also related to the size and structure of INGOs (Kriesberg, 1979). Noting the year of establishment of ISPAs compared to other INGOs will also suggest whether or not they have different periods of expansion. The data presented in Table 3 indicate that most ISPAs, like other INGOs, were established since World War II.

Size is an important organizational characteristic and has many implications, as we shall see, for other aspects of the role ISPAs play in the world community. The size of associations such as INGOs is a complex concept. Are we thinking of the number of persons who are directly members, of the number of persons who are represented by the constituent organizations, of the number of organizations which belong, of the number of countries represented in the membership, of the number of staff persons, or of the size of the budget? These all are measures of INGO size, and each may have somewhat different consequences for the structure and functions of the INGOs. We considered several measures in the analysis.

One important indicator of INGO size is the number of countries represented in the membership. ISPAs might be expected to differ from other INGOs in size of membership. On the one hand, professional and scientific associations would be expected to have universalistic membership criteria. Particularistic criteria of religion or political ideology would not bar membership. This should be reflected in representation from a

TABLE 4 INGO Categories by Number of Countries Represented

		INGO Category	
Number of Countries	*ISPA*	*Non-ISPA Occupational*	*Nonoccupational*
Fewer than 10	14.7	19.4	14.8
10 - 19	23.1	34.8	24.6
20 - 29	19.9	18.9	19.1
30 - 49	23.9	14.5	20.9
50 - 74	11.6	6.9	9.5
75 or more	6.8	5.4	11.1
Totals %	100.0	99.9	100.0
(N)	(921)	(592)	(325)

large number of countries. On the other hand, ISPAs are based upon great occupational specialization, and countries which do not have a highly differentiated economy would tend not to be represented. Consequently, the number of countries represented in ISPAs would be few. As can be seen in Table 4, ISPAs vary greatly in their extent of membership. In general, compared to Non-ISPA Occupational INGOs, they tend to have extensive membership. Over 40 percent of the ISPAs have 30 or more countries represented in their membership, compared to 27 percent of the Non-ISPA Occupational INGOs. Nonoccupational ISPAs have extensive memberships similar to the ISPAs, but they are even a little more likely to have very extensive memberships, 75 or more countries. Relatively many Non-ISPA Occupational INGOs have few countries represented because many of these INGOs are regional organizations.

One of the difficulties in comparing the size of different INGOs is their diverse types of members. Many INGOs consist of national associations, many of individual persons, some of corporate units such as institutes, and some of combinations of these member types. Inspection of Table 5 indicates that most INGOs which are not ISPAs have only associations as members. ISPAs are most likely to be composed solely of national associations although many have only individuals as members. Many other ISPAs have both associations and individuals as members. In making these groupings, we are ignoring whether the individuals are full or associate members. Non-ISPA Occupational INGOs are relatively likely to have corporate units as members.

Another measure of INGO size is the number of affiliated organizations. In some INGOs only one association from each country can belong

TABLE 5 INGO Categories by Type of Membership

		INGO Categories	
Membership Types	*ISPA*	*Non-ISPA Occupational*	*Nonoccupational*
Associations Only	34.5	51.7	52.9
Individuals Only	31.4	11.6	22.2
Corporate Bodies Only	5.7	13.3	3.8
Associations and Individuals	15.7	8.2	13.5
Individuals and Corporate Bodies	5.8	4.0	2.9
Associations and Corporate Bodies	2.5	6.4	2.6
All Three Types	4.4	4.8	2.0
Totals %	100.0	100.0	99.9
(N)	(1023)	(623)	(342)

and only national associations can be members; in those cases, the number of organizations would be the same as the number of countries represented. Some INGOs, however, allow more than one association from a country to belong; and in other cases, no associations belong. It turns out that the INGO categories differ in the number of organizations which belong about the same way as they do in the number of countries represented (see discussion of Table 4). In general, there is a shift toward more organizations than countries in each category of INGOs.

Budget size and number of staff are two important and traditional indicators of size. Unfortunately, this information is not available in the *Yearbook* for many INGOs. Furthermore, lack of information may or may not indicate the absence of a staff or a budget. Coding and interpreting the results, then, is fraught with difficulty. We will first consider the number of paid staff members. If we examine only the 58 percent of the INGOs with fairly complete entries in the *Yearbook* and assume that those which do not report any staff do not have any, almost two-thirds of the INGOs are without any paid staff. They may use volunteers, and a few use the staffs of other organizations. If we consider only the size of staff among INGOs which report some staff, in about one-third of the cases (38 percent), they have only 1 or 2 paid staff persons. Of the ISPAs, 23 percent have ten or more staff members, compared to 28 percent of the

Non-ISPA Occupational INGOs and 31 percent of the Nonoccupational INGOs. When we combine volunteer and paid staff, the results are similar.

Budget size is even more difficult to compare using *Yearbook* data. We will discuss only the approximately 400 INGOs for which we have budgetary information. First, the budgets of many INGOs are very small: about 20 percent report budgets of $5,000 or less. ISPAs tend to have modest-sized budgets, between $6,000 to $25,000, and the other INGOs tend to have somewhat larger budgets. Non-Occupational INGOs vary greatly in the size of their budgets, and more than 15 percent of them have budgets of $500,000 or more, compared to 14 percent of the Non-ISPA Occupational INGOs and 7 percent of the ISPAs.

In short, ISPAs tend to have relatively small budgets and staffs, even compared to other INGOs. This is probably related to the finding that ISPAs disproportionately have only individuals as members. In some ways, given the kinds of tasks they report conducting and the extensiveness of their membership, the relatively small budget and staff are particularly striking. All this may combine to affect the nature of the ISPA organizational structure.

Most INGOs, including ISPAs, have a simple structure of two levels: a general congress meeting and an executive council, committee, or board. ISPAs are slightly more likely to report only one level: that of the general congress. The other INGOs are slightly more likely to report having more than two organizational levels.

Even when ISPAs have more than one level, the smallest or highest level unit meets less frequently than does the highest level among other INGOs. Thus, almost one-third of the Nonoccupational INGOs have highest level units which meet two or more times a year: about 25 percent of the Non-ISPA Occupational INGOs meet this frequently, and only 13 percent of the ISPAs do. Similarly, the general membership congresses of ISPAs meet less frequently than do those of other INGOs. About 20 percent of the ISPAs have general membership meetings every four years or less frequently, compared to 11 percent of Non-ISPA Occupational INGOs and 14 percent of Nonoccupational INGOs. About 35 percent of the ISPAs meet as often as once a year, but 39 percent of the Nonoccupational INGOs and 54 percent of the other-occupational INGOs meet this frequently.

Interestingly, the pattern is different when we examine the number of committees among the three categories of INGOs. ISPAs are slightly more likely than other INGOs to have many committees. Other research (Kriesberg, 1974) has indicated that committees are alternative ways in which INGO members can conduct their activities. Perhaps having many committees compensates for the relative lack of hierarchical differentiation and of staff size.

The degree of activity, as measured by size of budget, staff, or frequency of meetings, indicates that ISPAs tend to be slightly less active than the other categories of INGOs. ISPAs are small-scale networks of individuals compared to other categories of INGOs. Of course, this does not mean that there are no ISPAs which are very large in scope and level of activity. We are only considering ISPAs, on the average, compared to other INGOs, on the average.

Finally, we compare INGOs with other INGOs in terms of their location within the world system. We consider the location of their headquarters and their relations with other international organizations.

Nearly all ISPAs, like other INGOs, have their headquarters in Western Europe, Canada, Australia, or the United States. About 90 percent of the ISPAs and the Non-ISPA Occupational INGOs are headquartered there and about 85 percent of the nonoccupational INGOs are. The Non-ISPA occupational INGOs are especially likely to have their headquarters in the United States; 12 percent do. This concentration of headquarters in the industrialized Western countries might reveal a parochialism which belies the global transnational impact of INGOs. It does indicate a concentration which reflects and reinforces the organizational density of the world. But the membership in INGOs is more extensive than the concentration of headquarters cities suggests. This was noted in the earlier discussion of the number of countries represented in the membership.

The location of INGOs in the world system can also be seen in the extent of their relations with each other and with international governmental organizations. A consultative relationship with one or more United Nations Specialized Agencies was the most frequently reported relationship. About 25 percent of the ISPAs report having a consultative relationship with UN organizations; other INGOs are slightly more likely to have such relations. Consultative relations with other international nongovernmental organizations are maintained by 11 percent of the ISPAs; again other INGOs are slightly more likely to have such relations.

On the whole, ISPAs vary greatly in their scale of activities, scope of membership, and involvement with other international organizations. Many ISPAs are relatively small, basically networks of colleagues working on shared intellectual interests. Others are large-scale federations of national associations intricately imbedded in an extensive network of international organizations. We turn now to examine these variations and their correlates.

ORGANIZATIONAL ASPECTS OF ISPAs

Given our interest in the social implications of ISPAs, we now focus our attention upon organizational aspects of ISPAs which might contribute to the development of a global society. We analyze variations in: (1) the level of ISPA activity, (2) the degree of cooperativeness in ISPA activities, (3) the frequency of membership meetings, and (4) the extent of linkages to other international organizations. Characteristics of ISPAs which might affect variations in these organizational aspects include ISPA goals, size, age, and type of membership.

To assess variations in the attributes of ISPAs, we will use the variables already discussed and combinations of them. The level of ISPA activity is indicated by the number of staff persons; the number is the sum of volunteer and paid staff members. Since information is lacking on many INGO characteristics, the absence of a specified number of staff members may mean that either there is no staff or that the information was simply not reported. Such cases were omitted for this analysis, and, therefore, we are considering only ISPAs with at least one volunteer or paid staff member.

The degree to which ISPA activities are cooperative is measured by an index based upon reporting or not reporting the following activities: engaging in joint tasks, providing services to members, and establishing common standards. The index number indicates that the ISPA reports engaging in none, one, two, or three of these activities.

The frequency of membership meetings is indicated simply by how often the general membership meets. Finally, the extent of linkages to other international organizations is measured by summing the consultative or membership relations with United Nations organizations, other international governmental organizations and other international nongovernmental organizations. Given the variety of types of relationships, an ISPA could have a score ranging from 0 to 13; in actuality, the highest score is 5.

Some, but not all, of these ISPA aspects are associated with each other. Thus, ISPAs which report carrying out a greater variety of cooperative activities tend to have larger staffs (Gamma = .25).[2] They are also somewhat more likely to have extensive relations with other international organizations. ISPAs with extensive international organizational relations are slightly more likely than other ISPAs to have large staffs and less likely to have frequent general membership meetings. There is no significant relationship between frequency of general membership meetings and (1) size of staff, or (2) the variety of cooperative tasks conducted.

In trying to account for variations among ISPAs, I will examine especially the following factors: the ISPA's age, size, sphere of activity, individual or association membership, and type of goals.

Age has been shown in earlier research to be related to various organizational developments (Kriesberg, 1979). If it turns out that ISPAs tend to become more active, engage in more cooperative activities, meet more frequently, and increase their linkages with other international organizations with age, this has implications for the role ISPAs may play in the development of a world society. Organizational size has been found to affect greatly many organizational characteristics, among INGOs as well as other kinds of organizations (Kriesberg, 1976; Blau and Schoenherr, 1971). In this analysis size is measured by number of countries represented. The number of organizations or individuals who belong (or are represented by member associations) is not available for many INGOs, and therefore we use this indirect measure. In this case, however, it does have substantive significance as an indicator of dispersion.

Third, we wish to learn whether particular professional and scientific fields are likely to affect the ISPA characteristics of interest here. We will look at sphere of activity as categorized in the *Yearbook of International Organizations.*

Another possibly important determinant of ISPA variations is whether individuals or associations are members. We will compare ISPAs with: (1) only individuals as members, (2) ISPAs with individuals and associations, and (3) ISPAs with associations or other corporate bodies or various combinations among them.

We also wish to know whether or not organizational goals affect ISPA variations; this has particular policy relevance, since goals are selected by organizational members. For purposes of analysis here, the ISPAs were divided into three categories in terms of goals: (1) those which seek to represent the interests of members vis-à-vis others or to promote conditions which serve member interests, (2) those which do not mention those aims, but mention serving nonmembers, and (3) those which do not mention any of those aims but refer only to facilitating the normal activities of the members.

These factors are generally independent of each other. The two which are most highly intercorrelated are age and number of countries represented in the ISPAs. Using ungrouped data, the Pearson correlation is .29 (N = 1,420). Among the ISPAs for which data are reported for size of staff, we find that staff is moderately correlated with number of countries represented. Using ungrouped data, the Pearson correlation is .36, and it is also correlated with age: r = .20 (N = 488). In this same sample, age is more highly correlated with number of countries represented than in the larger sample (Pearson correlation = .34; N = 488). These relationships among independent variables indicate that at times it will be necessary to examine how two or more independent variables affect organizational characteristics simultaneously.

We now turn to an analysis of the factors affecting the level of activities, the cooperative nature of tasks, the frequency of membership meetings, and the extent of international organizational linkages. As the indicator of level of activity, we use size of staff. First, we will note whether particular types of ISPAs tend to have very few staff members. On the whole, we do not find marked differences. About 45 percent of the ISPAs reporting any paid or voluntary staff have between 1 and 3 such staff members. ISPAs in the Agriculture, Commerce, and Economic spheres are especially likely to have few staff members. ISPAs in the Religion, Social Welfare, Transportation, and International Relations spheres are most likely to have more staff members.

The staff size is most clearly affected by the number of countries represented in the membership. ISPAs with many countries represented in their membership are likely to have many staff members. The age of the ISPAs is also related to staff size.

We also examined the possible effects of ISPA goals and type of membership. ISPAs with goals pertaining to serving member interests by representing them or by improving their conditions are no more likely than other ISPAs to have large staffs. There is only a slight relationship between type of membership and staff size. ISPAs with only individuals as members are especially likely to have very few staff members.

If we include all these variables in a single multiple regression, we obtain a multiple correlation of .37; nearly all of this is accounted for by the number of member countries. The standardized Beta for number of countries is .33. On the whole, these variables do not explain a great deal of the variance in the size of staff.

The second organizational aspect we shall examine is the variety of tasks of a cooperative nature conducted by ISPAs. Among the types of ISPAs, those in the spheres of Sports, Law, Social Welfare, Commerce, and Transportation are most likely to report undertaking cooperative tasks. The Health-related ISPAs are relatively low in this regard. Overall, the number of countries which are represented in the ISPAs tends to be most highly related to carrying out cooperative activities. (Using grouped data, the Gamma is .22; using ungrouped data, the correlation is .19.) Older ISPAs and those with associations and corporate bodies as members rather than only individuals tend to carry out cooperative tasks. Organizational goals, as measured here, are not related to the variety of cooperative tasks conducted. The multiple correlation of cooperative tasks with number of member countries, age, type of members, and organizational goals as independent variables is .22. The results of the analysis suggest that larger ISPAs, those with associational members, and those which are older tend to carry out more cooperative activities. But these factors do not account for a great deal of the variation in the extent of cooperative tasks carried out by ISPAs.

The third ISPA characteristic we are examining is the frequency of general membership meetings. Among the various types of ISPAs, those in the spheres of Religion, Social Welfare, Technology, and Education are most likely to meet infrequently; about one-third of them meet once every four years or more rarely. ISPAs in the areas of Transportation, Commerce, Bibliography, Economics and Finance, Art, and International Relations are most likely to meet as frequently as once a year or more often. ISPAs in the Social and Natural Sciences are most likely to meet once every three years (about 30 percent).

Variations in the frequency of general membership meetings appear to be unrelated to ISPA goals as measured here or to type of membership by individuals or associations. The number of countries represented and age are both related to frequency of membership meetings. Older organizations tend to meet more rarely, and ISPAs with more countries represented also tend to meet more infrequently.

One possibility is that as organizations age, they increase in size and in vertical differentiation. Such differentiation could result in organizational activities being conducted in specialized bodies and not in general membership meetings. It is true that there is a tendency for ISPAs with more levels to meet less frequently than do those with fewer organizational levels. But the relationship is not very strong. Furthermore, previous analysis of INGOs has found that age is not related to vertical differentiation (Kriesberg, 1976). In the present instance, we examined the relationship among ISPAs between the number of countries represented and the frequency of general membership meetings, holding constant number of levels. We found that ISPAs with a large number of member countries meet less frequently than other ISPAs, regardless of the number of organizational levels; but the number of levels is no longer consistently related to frequency of general membership meetings, controlling for number of countries represented.

The findings can be interpreted in other ways. Extensive membership is associated with adversary countries belonging to the same INGO and, in the case of U.S. and USSR comembership, this has meant relatively infrequent membership meetings (Kriesberg, 1974). Furthermore, the number of countries represented indicates the dispersion of membership. It may be that members who are widely dispersed are less ready to attend meetings frequently. Aside from this phenomenon, old ISPAs tend to schedule relatively infrequent general membership meetings. This may partly result from increasing age and utilizing other ways than meetings to conduct organizational affairs. It may also be a generational phenomenon. Thus, ISPAs formed in 1960 and more recently are likely to meet once every two years or more frequently (over 70 percent); of those established between 1930 and 1959, about 55 percent meet as frequently; while among those formed prior to 1930, fewer than 45 percent meet as often.

The final aspect of ISPA variation we wish to examine is the extent of relationships with other international organizations. First, we consider which types of ISPAs are most likely to have extensive linkages. We find that the linkages are most often extensive among ISPAs in the Social Science, Transportation, Health, and Commerce spheres. Such linkages are least likely to exist among ISPAs in the Religion field. As with other ISPA variations, the number of countries represented in the membership is strongly associated with the extent of relations with other international organizations. (Using grouped data, the Gamma is .32.) The age of the ISPA is also related to the extent of linkages: older ISPAs tend to have more extensive ties with international organizations. ISPAs formed in the decade of 1940-1949 are particularly likely to have extensive ties.

In addition to number of countries represented and age, the type of membership also affects the extent of linkages with other international organizations. ISPAs with only individuals as members are least likely to have any consulting or membership relations with other international organizations. ISPA goals have only a slight relationship to the extent of linkages. ISPAs which have as one of their aims helping nonmembers and do not also mention advancing the interests of their own members as an aim are most likely to have diverse ties with other international organizations.

In summary, level of activity (measured by staff size) among ISPAs is largely determined by the number of countries represented in the membership. In addition, being older and having associational or corporate members are also somewhat related to larger staff size. Carrying out activities involving cooperative effort is moderately related to the number of countries represented. In addition, having associations rather than individuals as members and being older are somewhat related to conducting cooperative tasks. Frequency of general membership meetings is also related to number of countries represented in the membership and to age, but inversely. Less dispersed and more recently established ISPAs meet most frequently.

ISPAs in different spheres of activities also vary in frequency of general membership meetings. The extent of linkages with international organizations is strongly related to the number of countries represented among the members. In addition, older ISPAs and those with associations as members tend to have extensive linkage. ISPAs in different spheres of activity also vary in their international ties.

Overall, the independent variables we have been examining do not account for a great deal of the variation in the characteristics of ISPAs which concern us. This may reflect the measurement problems we face. It may also indicate that a large number of factors interact to determine these characteristics. The role of idiosyncratic factors relating to the leadership and history of the ISPAs may also be important.

IMPLICATIONS FOR THE EMERGING WORLD SOCIETY

We now turn to consider some implications of these findings and information derived from interviews and other studies. What is of interest in this context are the possible consequences of ISPAs for the emerging world society and for the advancement of justice and the reduction of violence. Given the vast forces implicated in sustaining violence and injustice, we should hardly expect that ISPAs will have powerful effects. Yet, they can and do contribute to peace and justice, and I wish to discuss some of these influences and how the organizational characteristics analyzed earlier may aid these possible contributions.

Four possible effects of ISPAs could contribute to the reduction of violence and the attainment of justice. First, ISPAs can provide a setting within which nationals of different countries obtain insights into the policies and thinking of other countries and even develop mutually acceptable policies. These ideas can be transmitted to their home governments and can influence the governments to reach a mutually satisfactory solution of some intergovernmental problem. The Pugwash meetings contributed in this way to the development of nuclear testing agreements between the United States and the USSR.

Second, ISPAs can contribute to the successful carrying out of activities which benefit disadvantaged and deprived groups or help the scientific and professional members of the ISPAs in their work. In the first instance, for example, medical workers may help solve medical problems and provide medical service where there are special difficulties. In the second case, ISPAs help scientific and professional workers advance their knowledge and skills, e.g., when they help sponsor and arrange joint research and data gathering activities.

Third is the longer-run and more indirect consequences of ISPA activities: the development of a common world culture, at least in specific areas.

This may take the shape of common or shared ideas about social and natural phenomena. It may also be formed through increased awareness and tolerance of differences. Within the social sciences, for example, there may be increased familiarity with Marxist social thought by non-Marxists and a greater appreciation of non-Marxist social thought by Marxists as a result of participation in ISPAs.

Finally, the most extreme possible consequence is the development of alternative transnational structures through the vehicle of ISPAs. That is, ISPAs begin to provide quasi-governmental services on a nonnational basis. As more and more global organizations become alternative ways of providing services, the territorially-based state will become less significant (Lakey, 1973). Insofar as ISPAs carry out activities which bind their members in nomenclature, standards of procedure, and combined efforts, they are conducting governmental actions in the allocation of resources and determining how they should be collectively employed.

The variations among ISPAs in information and analysis, presented here, affect the impact ISPAs might have. As the analysis indicates, many ISPAs are small: they draw their membership from few countries and do not have associations as members. They do not have large budgets or staffs nor extensive links with other international organizations, and they do not carry out activities which require cooperative effort from members. Other ISPAs do have association members and have more extensive activities and membership. Some ISPAs are large: they have members from most countries of the world, are linked to other international organizations, have substantial staffs, and conduct activities which involve significant cooperative efforts. Let us think how these variations may be related to possible consequences of ISPAs.

First, consider the forum ISPAs may provide for exchanging information and insights and developing common solutions, any or all of which can be influentially transmitted to the nationals' home governments. It is probably true that the exchange of ideas is facilitated within a small association in which individuals have the least sense of *representing* a national organization, even a nongovernmental one. The staff or common cooperative activities would be irrelevant to such exchanges. However, when it comes to *influentially* transmitting the ideas acquired through such meetings, it may well be that ISPAs with associations as members are particularly effective, since persons from such associations would be especially credible and carry more authority with their own governments.

Furthermore, larger ISPAs probably carry more authority in the minds of government officials; and, therefore, their suggestions are more likely to be heard and given serious consideration. Conversely, large, important ISPAs may be worthwhile targets for governmental intervention. Secret government infiltration and support politicizes the ISPA in ways which are not easily recognized.

There are dilemmas, too, in connection with the second area of possible ISPA consequences: facilitating the work of members and their contribution to solving problems. Small ISPAs made up only of individuals are probably especially facilitative of mutual understandings about issues. The persons are likely to be self-selected in terms which are significant to the actions of the ISPA. But having associational members undoubtedly means much greater financial and personnel resources to carry out common endeavors and complicated tasks requiring coordination. Even the legitimacy needed to conduct comparative social research in different countries is undoubtedly aided by having a large ISPA with national associations from many countries constituting the ISPA. Thus, the International Sociological Association probably provides the vehicle to obtain widespread governmental cooperation for comparative studies which a group of individuals, even if they formed an international organization, could not provide. Our analysis indicated that ISPAs with association members were more likely to report carrying out cooperative activities than those with individuals as members.

The development of a world culture within and through the occupational communities represented by ISPAs also would appear to be facilitated by the interpersonal relations in which individuals interact solely as individuals.[3] But the self-selection which such ISPAs could make possible also limits the possible development of a culture shared by most members of the world occupational community. If national associations belong to an international organization, individual members of national associations who otherwise may have little interest in colleagues in other countries would be likely to pay some attention to their activities. Many persons who would not otherwise be aware of the work of their counterparts elsewhere in the world become acquainted with it. Of course, this may not result in harmony and mutual respect. New and realistic bases for disagreement and conflict will arise among people who previously had nothing but good will based upon ignorance. Relations may become politicized in the context of organizational struggles. Yet a shared culture must develop from the reality of differences as well as commonalities.

Finally, for ISPAs to become truly alternative structures to government agencies, they must have resources, a structure, and activities which bind members together. From our analysis, it seems that it is the large ISPAs with associations as members which are likely to have larger staffs and engage in cooperative activities. We have also seen that with time, ISPAs tend to move in those directions.

ISPAs can also perpetuate and reinforce national forms. What is probably needed are associations based not upon country but upon transnational identities within the scientific and professional spheres. Thus, there might be international working groups formed by persons working on particular intellectual issues, and these groups could be the building blocks for an ISPA. Several such ISPAs could federate to form an ISPA with more resources. These working groups could be fostered by a parent ISPA, or they might evolve from the ongoing activities of scientists and professionals. In either case, the working groups should be as extensive as possible, drawing upon all persons who share an interest in the subject matter.

ISPAs may not be the vehicles which in themselves transform the world. But they are part of the developing world society, and in many ways are in the forefront of the changes which are taking us away from a world divided into clearly bounded, impermeable nation-states.

NOTES

1. About 85 percent of all INGOs have persons or associations which belong in their capacity as incumbents or representatives of specific occupational roles. This number was calculated after omitting INGOs which lacked information to make possible classification in terms of members' occupational or nonoccupational roles.

2. For this and similar analyses, size of staff is grouped into five categories and tasks into four categories; N = 193; Kendalls Tau C = .17, Significance = .001, Chi Square = 18.93, Significance = .09.

3. ISPAs provide a setting in which acquaintanceships are made. These are often converted into personal friendships, collaborative colleagial relations, or consultative ties through correspondence, telephone calls, and personal visits.

REFERENCES

ANGELL, R. C. (1969) Peace on the March: Transnational Participation. New York: Litton.

BLAU, P. M. and R. SCHOENHERR (1971) The Structure of Organizations. New York: Basic Books.

EVAN, W. M. (1974) "Multinational corporations and international professional associations." Human Relations 27, 6: 587-625.

--- (1972) "MNCs and IPAs: an international organization research frontier." Int. Associations 24 (February): 90-100.

FELD, W. J. (1972) Non-Governmental Forces and World Politics: A Study of Business, Labor, and Political Groups. New York: Praeger.

KRIESBERG, L. (1979) "Age and organizational developments of international nongovernmental organizations," in R. P. Mohan (ed.) Management and Complex Organizations in Comparative Perspectives. Westport, CT: Greenwood Press.

——— (1974) "Organizational membership and structure: international nongovernmental organizations and co-membership from adversary nations." J. of Voluntary Action Research 3 (July-October): 34-40.

——— (1972) "International nongovernmental organizations and transnational integration." Int. Associations 24, 11: 520-525.

LAKEY, G. (1973) Strategy for a Living Revolution. San Francisco: W. H. Freeman.

SKJELSBAEK, K. (1972) "The growth of international nongovernmental organization in the twentieth century," pp. 70-92 in R. O. Keohane and J. S. Nye, Jr. (eds.) Transnational Relations and World Politics. Cambridge, MA: Harvard Univ. Press.

COMMENTARY

James R. Kurth

Do international scientific and professional associations (ISPAs) matter in the international system? In particular, have they made a difference in the policies of national governments, in the activities of international governmental organizations (IGOs), or in the activities of scientists and professionals themselves? These are important questions raised in this volume, and the two chapters by Louis Kriesberg and Diana Crane provide us with the beginnings of an answer.

The impact of ISPAs upon the policies of national governments. The most expansive claim about the role of ISPAs is that they have influenced the policies of nation-states. If so, this is one way in which ISPAs "can and do contribute to peace and justice," as Kriesberg states in his section "Implications for the Emerging World Society." And the most persuasive example, one mentioned by Kriesberg, is the Pugwash meetings, which contributed to the development of nuclear testing agreements between the United States and the Soviet Union.

But the papers by Kriesberg and Crane give no other examples of major policies of national governments in which ISPAs made a difference in the outcome (and technically, even the Pugwash meetings were not an example of an ISPA). There are many reasons, some given by Kriesberg and Crane, to believe that such examples are rare; the small budgets and small staffs of most ISPAs are an indication of their political weakness. If future research were to construct an inventory of cases in which ISPAs actually made a difference in national policies, it would be an essential and useful step in helping social scientists to specify the conditions under which ISPAs can and do have a policy impact.

The impact of ISPAs upon the activities of international governmental organizations (IGOs). Diana Crane's chapter presents a mapping and analysis of a network of international organizations which shows that much activity of major ISPAs is directed at and around a few IGOs, especially ECOSOC, UNESCO, and WHO. But, as she argues, the competition for influence among ISPAs could be so intense that in fact the ISPAs cancel each other out; she concludes in her section "Network Analysis As a Strategic Methodology," along with previous findings, that "individuals, experts, and special committees were more able to exert an influence upon IGOs than many INGOs" (including ISPAs).

Here again we need some empirical examples of cases where ISPAs made a difference in the activities of IGOs. Such examples are not given in the Kriesberg and Crane chapters.

Even if we were to find that ISPAs influenced the activities of an IGO, there is also the question about the importance of that IGO activity. Was that activity one which in turn had consequences for the policies of national governments and for the lives of human beings? Or was that activity merely yet another pompous resolution or dreary document, ignored by all governments and unknown to most persons?

The impact of ISPAs upon transnational relations. Another way in which ISPAs might have an impact upon the international system is as transnational actors, i.e., having a direct impact rather than one mediated through national governments or through IGOs. As Kriesberg argues in his section "Implications for the Emerging World Society," ISPAs could "begin to provide quasi-governmental services on a nonnational basis. . . . Insofar as ISPAs carry out activities which bind their members in nomenclature, standards of procedure, and combined efforts, they are conducting governmental actions in the allocation of resources and determining how they should be collectively employed." Regrettably, Kriesberg gives no example where this process has actually occurred, nor does he suggest the conditions under which it is likely to occur in the future.

Let us stipulate, however, that in the future, ISPAs could accumulate enough resources to become significant transnational actors. It would then be probable that the experience of ISPAs as transnational actors in, say, the 1980s would recapitulate the experience of multinational corporations as transnational actors in the 1970s. That is, the very growth in the power of transnational actors (multinational corporations) in the 1960s provoked a reaction by national governments in the 1970s (e.g., nationalizations or stricter regulations, especially in Latin America and in the Middle East). The end result was not the elimination of either transnational actor or national government but rather something of a symbiotic relationship between the two. But the national government normally has been able to impose important limits on the operations of the transnational actor. There is every reason to think that, if ISPAs were to become important transnational actors, national governments would impose comparable limits on their operations too. And since the resources available to the International Council of Scientific Unions (ICSU) will hardly match the resources available to Exxon, it seems that the limits imposed on ISPAs will be even more stringent than those imposed on MNCs.

The impact of ISPAs upon the activities of scientists and professionals. Finally, it has been argued that ISPAs nurture the development of scientific and professional disciplines. The chapter by Crane provides considerable evidence to support this proposition. If the study of ISPAs were limited to the study of their impact on scientists and professionals, it would be a modest enterprise indeed, essentially a case of professionals studying other professionals talking to other professionals. But at least the topic of study would be indisputably real and concrete.

My comments have suggested that the impact of ISPAs on the international system has been and will continue to be rather slight. There are, however, certain points in the policy process of national governments and of international governmental organizations where ISPAs might make a difference in the outcome. It often happens that, in the first phase of a policy process, the relevant issues are scientific or technical, rather than economic or political. Examples are the Research and Development phases of a weapons system, a space program, a new energy source, or a new telecommunications system. At this early point, economic and political forces (which can always overwhelm or distort scientific and professional forces in a direct conflict) have not yet been informed, energized, and mobilized around the emerging program. Scientific and professional forces have the field to themselves and can largely define the terms (and perhaps determine the outcome) of the later phases of the program and of the debates around it. It is at this early point that the information and influence provided by ISPAs could have a very substantial impact on the policy processes of national governments and of international governmental organizations. If so, then the greatest political victories of ISPAs would be those which would be virtually unknown to the world at large—and perhaps unknown to those social scientists who hope to discover the impact of ISPAs upon the international system.

COMMENTARY

Henry Teune

The primary question addressed in these two chapters is: What do they represent about the state of research on international nongovernmental organizations (INGOs) and, specifically, on international scientific and professional associations (ISPAs)? What we learn about INGOs from these two chapters can be broadly summarized:

(1) Historically, INGOs are a twentieth-century phenomenon;

(2) Since World War II, they have been growing exponentially (any social or political phenomenon that grows by about 10 percent a year for several years is presumptively theoretically interesting);

(3) ISPAs have grown faster in recent years than INGOs, but their growth now seems to be slackening; and

(4) There is such great variety of these legally recognized entities that the search for covariance patterns in their characteristics has so far yielded only a few low-level correlations.

INTERPRETATIONS AND CRITIQUE

The trend data on the growth of INGOs are sharper and more salient than any observed patterns among their organizational characteristics. Professor Kreisberg's chapter, a search for some kind of order in the phenomenon of ISPAs, demonstrates this. His primary variables are organizational—number of member countries, age of the organization, the presence of permanent staff, types of activity, and the subject matter or focus. There are few correlations, perhaps because of the lack of reliable data.

The concluding section on implications does not follow from these research findings. Professor Kreisberg speculates on some of the consequences, primarily desirable ones, of the presence and growth the ISPAs—reducing violence, helping disadvantaged scientific communities, and developing a common world culture. There are, however, several negative consequences of ISPAs, particularly on national scientific communities, especially those in developing countries. The internationalization of scientific activity that has spread so rapidly in developing countries has led, among other things, to dissatisfaction of scientists with their local research support, diminishing productivity; inappropriate research priorities, removing research from pressing local problems; and distortion of the development of research in the universities and institutes, splitting the research community into internationals and locals. Strains between local needs and international aspirations in some cases have justified political barriers to international collaboration.

Professor Crane attempts to cast INGOs and ISPAs into a general context. Her three approaches for understanding and explaining the growth of the ISPAs are functional—the consequences of the activity are used to explain those very activities. ISPAs exist to exert political influence; they grow because they service the professional interests of their members; they integrate activities of organizations, presumably to their mutual benefit. Such functional, explanatory approaches usually carry little information, especially since their negation makes little prima facie sense.

What is needed is a configuration of the mix of incentives for the formation, maintenance, and growth of these organizations. Clearly, such incentives exist, but they are a complex mix. Also, it is necessary to look at both individual and organizational incentives and the relationship between them.

Crane concludes her chapter by indicating the usefulness of network analysis for understanding the ISPAs. Network analysis can express complex structural relationships. Such relationships, however, need to be theoretically informed or else the yield will be a complex description with little understanding.

The explanatory approaches suggested by Professor Crane—integration, interest, and self-interest—might help explain why certain people participate in such organizations, but not why they have grown exponentially as a twentieth century phenomenon. To explain this it is necessary to have macrotheoretical approaches that will account for this striking historical phenomenon of the twentieth century.

EXPLANATORY ALTERNATIVES

Rather than looking at these organizations as organizations and considering their characteristics, their incentives, their functions, and activities, ISPAs must also be seen as transnational actors in an international system that itself is changing. Four approaches to the international system will be briefly presented.

GROWTH AND TECHNOLOGY

One redundantly observed characteristic of organized efforts to create and disseminate knowledge is that there is exponential growth, followed by peaking, and then a plateau. Such growth patterns have been observed for disciplines and fields within countries (de Solla Price, 1963). Once a discipline begins to grow, accelerating increases in growth occur in both people and communications. One explanatory interpretation of ISPAs as transnational actors is that their expansion is the same process as that of organized professional and scientific activity within countries. The foundation of the state and its growth in the Western countries was vitally dependent on structures of communications–canals, railroads, and roads, as well as mail and electronic communications. In the late nineteenth century, there were technological advances that could be used to further international communications, often marked by the establishment of the Universal Postal Union in 1875.

The growth of ISPAs can be interpreted as the growth pattern of scientific activities at the international level. If this is the case, then it should be possible to chart patterns of this growth for separate areas of professional and scientific activities from their point of origin to a point of projected leveling off. These different areas would have differential patterns of growth, depending on when they started, aggregating into a general exponential curve.

THE INTERNATIONAL POLITICAL SYSTEM

The major characteristic of the international system in the twentieth century, especially since World War II, has been the spread of independent state systems following decolonization to a point where almost every area of the globe is covered by independent, sovereign states. Such a political system or decentralized world order was largely the creation of the United Nations when it became clear that the great power system of world order had broken down after 1945.

The erection of state borders makes transnational activities difficult even in areas where cooperation is imperative. Each government, in a sense, is a barrier to international transactions. Thus, INGOs, growing more rapidly than IGOs, were invented and used as a means of transcending state boundaries.

If this explanation is appropriate, then the slope of the line of growth of INGOs as well as ISPAs should be a time-lagged function of the growth of countries in the twentieth century. Further, the exponential growth should be leveling off both because the increase in new countries now is practically nil, and because beginning around 1971 the United States and the Soviet Union, after it became clear that international order was breaking down into uncontrollable regional conflicts, attempted to replace the decentralized world order of Truman-Acheson with the great power system of detente.

CENTERS AND PERIPHERIES

A third explanatory approach using the international system as a point of departure is that the INGOs are the product of the major world power centers to relate and control their peripheries. If direct access is not possible because of the ubiquity of the state system, then organizational penetration becomes the alternative. Thus, the INGOs are the result of purposeful actions of great powers to relate to their peripheries. One aspect of this for INGOs is dependence relationships within them, including great power funding, higher participation of individuals from the centers than the peripheries, and control of headquarters and staff.

If this approach is valid, then we can account for the exponential growth of INGOs especially since World War II and can expect that exponential growth will continue in the near future. But, the extent to which this is true, will be primarily indicated by the pattern of participation in the INGOs. There should be clustering of activities, in particular INGOs with particular centers dominant—U.S., USSR, France, Japan, and so on with substantial membership from their peripheries.

GLOBAL DEVELOPMENT

INGOs can be seen as part of the long term process of the development of a global political economy (Morse, 1976). There is an underlying development dynamic that creates new actors, such as INGOs and new relationships, such as transnational activities (see Teune and Mlinar, 1978). INGOs are one element in the emergence of a global political economy.

If this approach is correct, then exponential growth should continue in the foreseeable future largely because of the positive relationship between level and rate of increase. But more importantly than continued quantitative growth in numbers, the INGOs and ISPAs should increase their members, expand their activities, involve people from more countries, establish permanent staffs, and begin to consolidate with other INGOs, both formally and informally.

CONCLUSION

A macrohistorical perspective on the international system must be part of any explanation of INGOs and their behavior. The consequences of these organizations will be guided by the theoretical approaches used to explain them. In the context of the global development of the international system, their impact would be profound. They would be increasingly significant actors in the transformation of the international system.

REFERENCES

DE SOLLA PRICE, D. J. (1963) Big Science, Little Science. New York: Columbia Univ. Press.

MORSE, E. C. (1976) Modernization and the Transformation of International Relations. New York: Macmillan.

TEUNE, H. and Z. MLINAR (1978) The Developmental Logic of Social Systems. Beverly Hills; Sage.

PART II:
ECONOMIC IMPACT OF ISPAs

4

NETWORKS OF ECONOMISTS: THEIR ROLE IN INTERNATIONAL MONETARY REFORMS

Harry G. Johnson

International monetary reform is necessarily a matter of international agreement of national governments, directly or through international monetary institutions (of which The International Monetary Fund is the foremost and most formally organized). For obvious reasons, neither the bureaucracy of the IMF and similar organizations, nor the economic civil servants of national governments, can create or constitute a "network"; while international networks of professional economists can be formed, their role is necessarily informal and "consultative." This chapter is concerned primarily with description, secondarily with tentative analysis, of two sorts of such "networks" that came into existence in the past fifteen years or less. The first is the network that came to be known as the "Bellagio group," which began in 1963, and which was directly concerned with discussions of international monetary developments in the specific framework of reform of the international monetary system. The second is the series of closely overlapping international networks of academic economists that began with the Chicago Conference on the International Monetary System of September 1966, which was concerned with theoretical problems of the international monetary system. That conference fairly directly represented the dissatisfaction of R. A. Mundell with the method of proceeding of the Bellagio group. In turn, it led both to a series of conferences held in various European countries, sometimes organized around the central theme of the international monetary system and sometimes around domestic monetary theory, policy analysis, and empirical testing with the international aspect playing a subordinate part; and to major research enterprises in England (The University of Manchester and the London School of Economics), Switzerland (The Graduate Institute of International Studies, Geneva), France (The University of Paris–IX Dauphine), and the United States (the University of Chicago, and the University of Rochester).

THE BELLAGIO GROUP

What came to be known as the "Bellagio group" began in the autumn of 1963, after the annual meetings of the IMF and the World Bank at which the United States announced a switch in policy from previous denial that any reform of the International Monetary Fund was necessary to endorsement of the desirability of reform, and the establishment by "The Group of Ten" (major countries in the international monetary system) of Working Party Three of the Organization for Economic Cooperation and Development to study and report before the next IMF meeting on a plan for international monetary reform. Under-Secretary of the Treasury Robert V. Roosa, in a press conference, was asked why there were no professional economists in the group; his reply (paraphrased) was "because professional economists never agree with each other." This stimulated Professor Fritz Machlup of Princeton, famous for his interests in methodology and semantics, to set up and obtain foundation support for a series of meetings of economists to discuss and determine the reasons for their disagreements, on the methodological principle that disagreement must be the result either of differences in empirical assumptions or errors of analytical logic—both corrigible by scholarly research and discussion—or of differences in value judgments, which needed to be recognized and clearly stated.

The key figures in the organization and management of the group were Professor Machlup, Professor Robert Triffin of Yale, Professor William Fellner of Yale, and Professor Gottfried Haberler of Harvard, who took turns in providing a university umbrella for obtaining and administering foundation support, and in acting as chairman of the group. Other memorable participants were Sir Roy Harrod of Oxford, Jacques Rueff of France, Robert Mundell (initially Canada), Peter Kenen, and Richard Cooper (both United States). Fellner, Haberler, and Machlup themselves belong to a network, the by-then long-scattered products of the "Vienna" or "Austrian" School. Triffin, a Belgian by birth, had played and continued to play an important role as an expert and consultant on, and deviser of plans for, international monetary arrangements, and had moreover the priority of having been the first (and effectively the only) international monetary expert to call attention to the parallel between the IMF system and the interwar gold exchange standard, and the virtual certainty that the IMF system would break down for the same reasons. In this connection, it was relevant to the evaluation of the group that Triffin was at that time involved in personal rivalry for international monetary influence with Edward Bernstein, the first and long-tenured research director of the International Monetary Fund. In consequence, his initial activites in the group were counterproductive, in the sense that his aim was to induce the group to endorse collectively and publicly a plan designed by himself, and for this purpose he deployed all the skills of the clever committeeperson, in particular that of trying to focus attention on the precise wording of a draft document, and being extremely flexible in responding to criticisms by producing revised draft sections, as a means of preventing discussion of whether the purpose of the group was best served by producing a committee report—which in this case it definitely was not. The third meeting of the group, held in Princeton in March 1964, saw the high tide of this endeavor, with the group spending much of its time in arguing about a draft document. The result was widespread dissatisfaction, particularly among the younger participants, which might have terminated the enterprise had it not been that the fourth meeting was scheduled for the Villa Serbelloni in Bellagio, very attractive bait for participants, and that by then Professor Machlup as chairman had taken firm charge of the proceedings.

Machlup, though he was an expert chairman, had some irritating characteristics of his own. In the first place, he was apparently convinced that fulfillment of the obligation to the financial supporters required both a full week of meetings with most of the waking hours programmed. This meant very little if any time for free socializing among participants, especially as the Villa Serbelloni's catering arrangements were themselves both very formal and tightly time-tabled. On the other hand, it is probable that a looser time table would have resulted in the common tendency of such groups to break down into national or linguistic subgroupings without any enduring formation of international connections. Secondly, Machlup as editor of the influential Princeton Series of Essays in International Finance was determined to produce some sort of publication out of the group's deliberations (apart from probable obligation to the financial sponsor), and to do so before the September IMF meetings. The result was that the Bellagio meetings of January and June 1964 were very hard work, especially the former, with groups of two or three desperately putting together drafts of sections of the final document, in the short intervals between formal sessions and formal mealtimes and after the formal evening meetings ended. This effort, however, also probably contributed to the development of durability in the network of international contacts.

The final publication attracted considerable academic and financial attention;[1] and there was a substantial sum left over from the original grant. Machlup proposed to use the remaining funds for a joint meeting with the officials negotiating in Working Party Three and this suggestion was favorably received and acted on. The first joint conference of "the Bellagio group" proper occurred in January 1965, and was successful and interesting enough for the officials to want to repeat it. The result was a series of meetings every six months or so, allied with informal short meetings at the annual IMF meetings of those who were attending (the Fund regularly invited a selection of academics who could usually find their own funds, though some came with funding from their financial consulting clientele or from journalistic sources).

The early meetings, particularly the first, were formal and stilted. On the one hand, Machlup was extremely deferential to the officials, among other things summoning the academics for two days in advance to prepare an agenda which the officials tore up as soon as they arrived, and expecting them to stay a couple of days after the officials had left (they attended for at most two days and a half, usually effectively only for a day or a day and a half). These wasteful time-tabling practices were gradually modified, especially the agenda preparation, though they were never scrapped. Gradually, also, the officials began to be treated with deference adjusted to their revealed intellectual capacity. On the other hand, the officials, for whose convenience these meetings were planned to follow immediately after the meetings (often bitterly denunciatory, apparently) of Working Party Three, at first were very much on their guard as representatives of their respective nations. This continued to be the case with the British official representatives, who, whether or not they had the intellectual resources for intelligent discussion of system issues, invariably "stone-walled" on issues of both British financial policy and of other countries' policies, and as a result came to be treated with politeness but no great respect. Some of the American officials, especially the higher-placed ones, were no better, but their views were backed by the power of action and hence treated respectfully. The Canadians had no ideas, and gradually dropped out; the Japanese may have had no ideas either, but their silence was accepted as related to their language problem, and when maneuvered into speaking they performed creditably. The French were aloof but clearly of high intellectual quality; the Italians were also impressive. Gradually, two personalities emerged as dominant, in terms of intellectual power and toughness—Jonkheer van Lennep of The Netherlands and Otmar Emminger of the Deutsche Bundesbank, van Lennep being the tougher and Emminger more clearly the lucid intellectual. The proceedings tended to be shaped by the most recent international monetary crisis or flurry, and to begin with a lengthy theoretical exposition by Emminger, more rarely van Lennep, go through a strong criticism and weak defense of the most recent culprit—usually the British, occasionally the Italians—and wind up with a statement, respectfully heard and discussed but not always very respected, of the American position from the top American present. (In this connection, the Americans did typically send their best people, as did the other countries with the exception of the British.)

As it turned out, in my judgment the main educational effect was on the academics rather than on the officials. The officials' thinking developed faster and more boldly than that of the academics, presumably partly because they were continually discussing new crises and integrating them into their thinking, in the concrete sense that the officials penetrated the theoretical core of the problem of creating new and genuinely international liquidity, and embodied their understanding in the device of Special Drawing Rights at the IMF, while the academics were still laboring with the conventional concept of an international central bank. To explain this point, conventional concepts of banking, both commercial and central, entail the issuance of liabilities (money) against assets in the form of a mixture of reserves and of "sound" securities. For a world central bank, however, there is no need for reserves, since there is no one who can demand conversion of liabilities into alternative forms of wealth; and no need for "sound" securities either, since the point of holding securities is the ability of the bank to convert them into reserves to meet demands for conversion of liabilities into such reserves. What is needed, instead, is agreement to accept such international money in payment of international deficits, together with an obligation on holders not simply to spend their initial stocks to exhaustion but to run their affairs so as to maintain their stocks on the average, while running them up and down as short-run developments require. These two obligations—to accept the money in exchange, and not to spend it once-for-all—in turn require that the stock of such money is correct in quantity and grows at the rate appropriate to balance demand and supply without exerting either inflationary or deflationary pressure on the world economy. The Special Drawing Rights plan embodied these understandings, though the detail was still cluttered by the wreckage of past ideas and bargaining positions, and unfortunately the plan was implemented just in time to add fuel to the rapidly-developing world inflation.

It may be that the Bellagio group discussions were more influential than suggested, in promoting the thinking of the officials and leading them to think about the system as a whole. At worst, it provided them with an intelligent audience at which to project their own thinking and on which to test out which ideas were national-interest-oriented and which had general validity. At best, it stimulated them to think in system terms and to think theoretically rather than in terms of practical day-to-day operating problems.

The academics, as stated, seemed to me to derive much clearer benefits. One very practical one was that almost all of them found subjects to write about, subjects whose academic importance was established by involvement in the international monetary decision-making process and guaranteed publishability in the important (and incidentally lucrative) bank journals, and sometimes in the regular academic journals, not to speak of annual meetings sessions and conference sessions and volumes. Moreover, after a meeting there was always a new set of problems, permitting the writing of a series of continuing papers of the sort required to establish a man as an acknowledged professional expert in his field. At a more academic and scholarly level, the academics were forced to abandon many of the national biases in what they had previously believed to be objective analysis of the international monetary problem—especially American chauvinism with respect to the role of the U.S. dollar and contempt for the European proponents of the gold standard. There was in addition a trend towards acceptance of the need for greater exchange rate flexibility, even among the firmest defenders of the IMF system of pegged but alterable rates. At the professional level, there were two sorts of consequences. One was the development within the national groups of academic economists of national subnetworks in relation to the rest of the national profession, evident in the United States in a willingness to co-sign letters to The New York *Times* without the usual self-paralyzing academic bickering over the precise wording of the message to the nonacademic public. The other was the tacit establishment of a more or less agreed ranking of international monetary economists by academic quality, cutting across national lines with little or no reference to nationality, a ranking which transformed the standard of judgment from a national to an international one. This ranking persisted in the thinking of the participants in other contexts, with the result that, apart from mathematical economics and econometrics (for obvious reasons), the international monetary specialization is the most international of the many fields of economic specialization. (This characteristic, it should be noted, distinguishes international monetary economics from the sibling field of international trade theory, where the rankings are far more confined within national professions, and national eminence is achievable with scholarly achievement varying from the highly competent to the completely scientifically illiterate but politically clever.)

This particular by-product of the Bellagio group network was, however, more the result of the second type of network formation, to be discussed later, than of the Bellagio group itself. In addition the result, so far as it can be credited to the Bellagio group, had a great deal to do with the composition of the group itself, and particularly with the predominantly mixed backgrounds of the members. Specifically, the "Austrian group" was by no means unique in having a country of citizenship different from the country of birth and education, while others had different countries of citizenship and university employment and residence; and among the single birth-citizenship-residence category, some had worked long abroad or in various international activities. Hence there was an initial cosmopolitanism of outlook, together (and more subtly) with a concern for universal theory rather than national-interest-oriented theory, than would be likely with a group led by and composed of national professional leaders.

The last meeting of the Bellagio group that I personally attended was held in January or February 1973, just before the IMF system collapsed into the system of "dirty" or "managed" floating exchange rates. One further fact about it remains to be mentioned. During the Kennedy and Johnson administrations, devotion to the immutability of the dollar price of gold was so strong that it was impossible for the subject of floating exchange rates, or even of greater flexibility of rates, to be discussed in the Bellagio group. (The most one could discuss was devaluation or appreciation of a currency other than the dollar, and the officials invariably favored monetary and fiscal policy, sometimes trade liberalization, over exchange rate adjustment.) Since the three "Austrians" and several others were long-time firm advocates of floating rates, they organized a series of conferences on alternative exchange rate systems, with foreign exchange market dealers and commercial bankers taking the place of Treasury and Central Bank officials. This series, culminating in the publication of *The Burgenstock Papers* (Halm, 1970), was far more educational for the nonacademic contingent, though it also was somewhat depressing for its lavishing of academic time on the "wider band" and the "crawling peg" schemes. (The wider band means a wider margin of fluctuation for market exchange rates about the official parity price in terms of gold; the crawling peg means that the parity value is adjusted slowly to the market pressures by being automatically adjusted to some past average of market rates.) The educational effect was apparent in the rapid adjustment of market "practitioners" to the idea that a floating rate system could work, once the officially unthinkable was declared with academic authority to be in fact thinkable, and experience showed it to be highly profitable as well.

THE "PROBLEMS" APPROACH TO THE PROBLEMS OF THE INTERNATIONAL MONETARY SYSTEM

The second, and more lasting and important, type of network that has developed in relation to the problems of the international monetary system owes its origin to R. A. Mundell, who organized the 1966 Chicago Conference on Monetary Problems of the International Economy (Mundell and Swoboda, 1969).[2] Mundell was a brilliantly original international monetary theorist in his own right, and a brilliantly original academic entrepreneur. In addition, he had a knack for finding promising younger economists who could pick up and develop his ideas. The Chicago Conference departed sharply in intellectual style from the pattern of Bellagio, in emphasizing what came to be known as "the problems approach" to the international monetary system, as contrasted with the "plan" or "reform" approach. It departed sharply also in membership and responsibility structure, with the "elder statesman" being promoted to the position of session chairman and the papers being written and officially discussed by younger economists. The result was that the papers incorporated much more serious new thinking than is characteristic of a conference dominated by seniority (the worst in economics are the International Economic Association's quadrennial Congresses, which are dominated by national professional leaders, especially now that the "socialist" countries participate, demand equal time with the West, and send massive delegations at public expense). More fundamentally, the format and the "problems approach" gave the young economists the chance to shine by virtue of technical theoretical competence, which they usually possess in greater quantity and quality than their seniors, rather than to be cowed into respectful silence by virtue of the inferiority of their experience, wisdom, and public respect by comparison with their elders. At the same time, the young are generally harsher judges of their contemporaries' technical competence, and less confident of their innate national superiority over the lesser breeds of foreigners, than are their elders, so that international relationships based on mutual recognition of professional interest and ability are more likely to develop among them. For these reasons, the Chicago Conference both established a network of young international monetary specialists crossing national borders without much impediment, and permitted the development of further subnetworks as the individuals present at Chicago imitated the Mundell method of conference-convening as a means of institution and influence building within national professions and used their international connections to elevate the status of their conferences and to build up their influence and the status of their colleagues. Thus overlapping networks were created centering on The University of Paris–Dauphine, The Graduate Institute of International Studies in Geneva, the Graduate School of Business at Chicago, and somewhat later at the Institute of International Economic Studies in Stockholm. These networks were extended and reinforced by ad hoc coalition with two newly emerging networks concerned primarily with domestic monetary problems. One was the Money Study Group in the United Kingdom, a group organized on a national basis with two objectives in mind, both connected with the stimulation of British monetary research: development of contacts among academic economists, commercial bank, and other financial center economists, the Bank of England staff and Treasury monetary specialists, groups which until then had had few contacts and much ostensible contempt for one another; and organization of a "monetarist" or "quantity-theory" approach to money and monetary policy in rivalry with the stultifyingly deeply entrenched Oxford-Cambridge Keynesian approach to macroeconomic policy. The other was the Konstanz Seminar, an annual event organized by Karl Brunner as a means of propagating the views of himself and Allan Meltzer against the dominating approach of the European central bankers and monetary managers. Both the Money Study Group and the Konstanz Seminar naturally had to deal with international monetary problems as well as domestic (in sharp contrast to the Friedman, Brunner-Meltzer, and related quantity-theory approaches in the United States, which could be and was treated as effectively a closed economy), and to do so naturally established bridging networks with the international-money-oriented groups on the Continent and the networks linking them. In addition, the academic network system based on European universities was reinforced by the continuing presence of Mundell, in his capacity as owner of an estate near Siena, where, with support from a wealthy local bank, he continued to run annual international monetary conferences, though his interests shifted noticeably from an emphasis on academic research and theory to an emphasis on influencing international monetary opinion; and his clientele shifted accordingly from academics to international commercial banking leaders.

The various interlocking networks just described have had, in my judgment, a potent influence on the development of at least one section of the economics profession in Europe, and in particular had a powerful effect in both raising the scientific level of understanding, analyzing, and writing about international monetary affairs, and in stimulating the development of a European- and world-oriented profession in replacement of the tradition of noncompeting national professions. Not all of this development, of course, has been unqualifiedly to the good: in particular, some aspects of Mundell's early work of the question of optimal currency areas, in his own later writings and those of some of his European disciples, have been perverted to the service of an anti-American European nationalism of a somewhat repulsive kind.

On the positive side, however, the networks can be credited with responsibility for a considerable, and very rapidly achieved, improvement in understanding of both the problems that led to the breakdown of the IMF system, and the problems of the current system of managed floating rates. Under the former heading is the recognition of the reality of world inflation in a system of fixed exchange rates, and its basic origins in excessive monetary expansion, particularly in the United States as reserve currency country of the system. This concept was developed and emphasized in the course of a panel discussion at the University of Paris–Dauphine, and became the object of a major research program conducted at the University of Manchester, with network connections with related programs on a smaller scale at the London School of Economics and at the Graduate Institute of International Studies, Geneva. These programs were initiated several years before The Brookings Institution inaugurated a large program on world inflation, and by then had achieved a far higher standard of both theory and econometric application than was available in the state of the art in the United States. As is unfortunately often true of research programs in these days of dependence on foundation support, however, the programs have been restricted by the terms of their grants to study of the international monetary system before the collapse of the fixed exchange rate regime, and unable to shift as extensively and rapidly as would have been desirable to the problems of the new regime of floating rates. New theoretical and empirical research on the new problems has tended to be led by economists located elsewhere in the system of networks, notably—but by no means exclusively—economists at, or connected originally with, the University of Chicago.[3] The existence of the networks does, however, have the effect of keeping the international monetary economists in the various centers in close touch with the initiation of the newer as well as the completion of the older lines of research. In consequence, it is extremely unlikely that there will be a recurrence of the extreme kinds of difference among economists, both intranationally and internationally, that were the occasion for the initial convening of the Bellagio group. For those differences derived very largely from a combination of a predilection for pure theoretical model building, without reference to the need for rigorous empirical testing of theory against facts (which in turn requires both operational formulation of theory and attention to such factors as economic and political expectations and time-lags in economic responses neglected in traditional economic theory), with an intuitive and personally-experienced view of the world and how it operates. The difference can be encapsulated in the proposition that, whereas the older generation of economists is inclined to say "the floating rate system does not work the way I expected, therefore the theory is wrong, the world is irrational, and we can only retire rationality by returning to some sort of fixed rate system to be achieved by cooperation in management among national governments," the younger generation is inclined instead to say "the floating rate system is a system that should be expected to operate rationally, like most markets; if it does not seem to work rationally by my standards, my understanding of how it ought to work is probably defective, and I must work harder at the theory of rational maximizing behavior and the empirical consequences of it in the behavior of a floating rate system if I am to achieve understanding." This latter approach is the one that is being disseminated, and intellectually enforced, through the networks. Whether it will alter the official approach to the international monetary system, and in particular control the virtually automatic assumption that if the system does not work the way that practical men of affairs and government expect it to, the only solution lies in government intervention in the determination of exchange rates, is of course a different question.

CONCLUDING OBSERVATIONS

This chapter has been concerned with a description and evaluation of the informal networks of professional economists that have developed since 1963 in connection with the problems of the international monetary system and the reform of that system. It is extremely difficult to evaluate the relative influences of the personalities involved and their backgrounds, the nature of the problem and its setting, and the contemporary state of professional work in the field, in determining the activities of the network, and even more difficult to evaluate its net effect on the development of professional thinking, research and teaching in the area of study with which the network is concerned. What has been presented here is a personal impression rather than a scientific study. It is hoped, however, that the chapter will serve to suggest some ideas and raise some questions about the role of networks in influencing research and thinking in economics, and more generally in the social sciences. In this connection, two generalizations may be ventured. First, the Bellagio group had far less significant and lasting influence on the international monetary economics academic profession than the Mundell groups, largely because the former did not and the latter did bring modern monetary theory and mathematical and statistical methods to bear on the problem, and suggest fundamentally new research approaches and problems. Second, the Bellagio group—the only one focussed on policy and in contact with high-level policy-making—had little if any influence on the policy decisions, except in the general sense of providing a sounding-board for the policy makers. What it did do, and this could have been very important, was to educate the academics on the issues, and thereby ensure a reasoned reception for the policy decisions by a group of writers in influential journals of financial news and commentary.

NOTES

1. Chapter 1 of Machlup, 1964 contains a full account of the proceedings of the group.

2. Dr. A. R. Nobay has made in private discussion the important point that the dissatisfaction with the Bellagio group approach that led Mundell, supported by Johnson, to "the problems approach" was the consequence of Mundell and Johnson both being highly trained in both "real" and "monetary" trade theory, and aware of the need for an integration of these two aspects of international economic theory in a general equilibrium framework, in contrast to the tendency of the "plan" exponents to rely on ad hoc, largely "real," partial equilibrium theory; and that this fact underlies both the dominance of Chicago in subsequent developments and the rapid development of a distinctive theoretical monetary-general-equilibrium approach by the new generation of young international trade theorists.

3. See in particular the papers, and their authors, in International Economic Studies, 1976.

REFERENCES

HALM, G. N. [ed.] (1970) Approaches to Greater Flexibility of Exchange Rates: The Burgenstock Papers. Princeton, NJ: Princeton Univ. Press.

Institute for International Economic Studies (1976) "Proceedings of a conference on 'Flexible Exchange Rates and Stabilization Policy'." Scandinavian J. of Economics 78, 2.

MACHLUP, F. [ed.] (1964) International Monetary Arrangements: The Problem of Choice. Princeton, NJ: Princeton International Finance Section, Princeton Univ. Press.

MUNDELL, R. A. and A. K. SWOBODA [eds.] (1969) Monetary Problems of the International Economy. Chicago: Univ. of Chicago Press.

5

PROJECT LINK: POLICY IMPLICATIONS FOR THE WORLD ECONOMY

Lawrence R. Klein

The central research strategy of the project on the international linkage of national economic models (LINK) is to tie together, in a consistent way, the ongoing major macroeconometric models being used in each of the main countries or regions of the world, on the assumption that the model builders of each country know their own countries (area) best. An alternative approach would be to construct a standardized (perhaps simplified) model for each country or region, centralized in one research institute, and link them together by a given mechanism.

An American model for the United States, a Canadian model for Canada, a Japanese model for Japan, and so forth, have formed the constituent parts of LINK. The approach is so general that we have allowed some models to be annual and some quarterly; some to have large monetary sectors and some mainly real; some are demand-oriented and some are supply-oriented.

THE ORIGINS OF PROJECT LINK

LINK grew out of discussions in 1967-1968 in the Committee on Economic Stability of the Social Science Research Council. After looking at problems of domestic stability, it was decided that we should spend some time examining the international transmission of economic fluctuations or economic disturbances. So, we turned our attention to international stabilization. We brought together existing teams of model builders in various key countries and have since gradually expanded that group. LINK is an attempt to link together, by uniform specifications through the trade accounts of different countries, the existing econometric models now in use in all the major countries of the world. LINK, then, is virtually what is given in the title of a book edited by R. J. Ball: *The International Linkage of National Economic Models.* R. A. Gordon, Bert Hickman, Rudolph Rhomberg, and I were coprincipal investigators making up an executive committee in the early efforts of the Project.

The Committee on Economic Stability (now Stability and Growth) of the Social Science Research Council had some background experience in the fostering of cooperative research ventures since its founding in 1959. It sponsored the Brookings-SSRC econometric model project, combining expert (sector) input from all over the nation. It later sponsored the collaborative work on the economics of the monetary mechanism that eventually became the MIT-PENN-SSRC Model built for the Federal Reserve Board.

At a meeting in London in 1967, sponsored by the SSRC committee on "Is the Business Cycle Obsolete?", we took an outward looking international view towards economic stabilization policies. At a later committee meeting, we decided to look into the workings of the international transmission mechanism, i.e., the transmission of business cycle developments. We assembled a small international working party (United States, Canada, United Kingdom, Western Europe, Japan) in Stanford in summer 1968 and decided to go forward with a formal research project, supported jointly by the International Monetary Fund and the National Science Foundation.

We took the view that the national model builders and proprietors each knew their own countries (economies) best and used only existing national models with all their diversity of structure, but tied them together with uniformity on the trade side.

The project, called LINK, has always functioned as a loose cooperative research venture, guided by an executive committee from the SSRC Committee on Economic Stability. Modest research sums (about $7000 annually) were distributed to international research centers in order to support their linkage research, but the coordinating center was placed at the University of Pennsylvania.

The early concept included only the main industrial countries, but Project LINK had to initiate model building efforts for two major countries, Italy and Japan, where ongoing projects could not be found. From the beginning, UNCTAD represented the interests of the developing countries, and the centrally planned economies. Gradually, the latter interest has shifted to UNCDPPP. From time to time, socialist countries and some developing countries have sent representatives to LINK meetings. They are held twice yearly, in spring and late summer (early fall), rotating to centers around the world.

The membership-participation has grown. Now there are 13 separate country models, models of four developing regions, and models of the centrally planned economies in CMEA. As many as 40-50 people regularly attend LINK meetings. These meetings deal with world economic forecasts, international economic policy, and techniques of world model building. The participants come from four kinds of institutions—academic centers, research institutions, national government offices, or international agencies. We are held together in this loose confederation by some modest financial disbursements to participants, but more firmly by keen interest in the methods and substantive results.

OVERVIEW OF THE LINK MODEL

When individual model building teams in different countries apply their systems to problems of forecasting policy formation, or scenario analysis, they customarily assume exports to be exogenously given or mainly determined by other factors that are exogenous. The same is true of import prices, which depend mainly on pricing decisions by partner countries. By contrast, imports and export prices are endogenous variables for individual country models. To a large extent, imports depend on domestic activity levels, and export prices depend on internal costs.

The principal purpose of the LINK model is to endogenize exports and import prices on a world scale. It not only endogenizes these two types of magnitudes for all LINK participants together; it does so in a consistent way. Exports are the same thing as imports by partner countries. Similarly, an index of import prices is an average of export prices of partner countries. Consistency is achieved by our building a restriction into the world (LINK) model such that

$$\text{total world exports} = \text{total world imports.}$$

This basic accounting identity—law of conservation, if you will—is imposed in both nominal and real terms, i.e., in both current and constant prices.

Given the heterogeneity of individual country models—by design, according to the principle that resident experts will know best how to model and interpret the economic relationships of their own countries—some strict rules have to be imposed in order to implement the accounting restrictions. The rules concern valuation and classification.

In order to add imports and exports across countries to reach world totals for aggregate trade, a common unit of measurement has to be used. Each model, except for the regional models of developing countries, is denominated in local currency units. For world trade calculations in LINK, all export/import data, together with their corresponding prices, have to be converted through special linkage equations into U.S. dollar units. Another numéraire could have been used, but the U.S. dollar is the most frequently used unit of international account.

Numéraire units are not enough. Foreign trade magnitudes are valued at different points, inclusive of different charges. It is necessary to value both exports and imports on an FOB basis, at the point of initial export shipment. Many countries value imports CIF. Where this is done, we have reduced CIF values to an FOB basis. Service or "invisible" items that account for FOB/CIF differences are separately modeled in each case.

The consistency calculations hold for countries' total trade but also for their trade in various subclassifications. The LINK Project has subdivided trade into standardized classes for

- food, beverages, and tobacco;
- basic materials;
- fuels; and
- manufactures and semimanufactures.

In order to appreciate and analyze major developments in world agricultural markets, metal markets, and oil markets, it has been very important to have these separate classifications in LINK analyses.

This exposition gives only a broad overview of the process of consistent world scale linkage. More details are given in the Technical Appendix.

SOME LINK APPLICATIONS

The most straightforward case of LINK system application is simulation of the model—either over an historical period or into the future in genuine forecast applications. Both applications have been made extensively but it may be more interesting to explain the procedure of forecast simulation into the unknown future. Customarily, the model has been projected three years into the future, but system simulations have been made for longer durations. There is no reason to stop at the three-year forecast horizon, other than the amount of work to be done in assembling input values; accordingly, a "stretched" simulation for another 3-8 years is described in a subsequent section.

Toward the end of a calendar year, all LINK model proprietors are requested to send new input values for the year just completed (not fully known in economic reporting detail at the time) and for three future years. The inputs consist of initial conditions (lagged values of variables), future values of exogenous variables, and any equation changes since the last LINK solution. They are sent by card deck, tape, or telex listing of input changes. Every separate model is solved in pre-LINK mode at LINK Central (Philadelphia) to see if the model proprietor's solution can be duplicated. A complete listing of pre-LINK solutions is presented to all participants, at an international meeting in March. There may also be a first attempt at full post-LINK simulations. The main point is that all model proprietors see the whole range of country/area results all at once. After the March meeting, the post-LINK simulations are revised with reconsidered input values and adjustment to new readings on the economy. Another round of forecast critique occurs at the end-of-summer meeting (August-September) and a final post-LINK simulation is issued during the closing months of the year. At that time, a new three-year forecast cycle begins.

International trade data are reported with considerable time delay; therefore, at the time the first calculations are being made in a forecast round for years t, t+1, t+2, only scattered information is available for t-1. The first step in a LINK forecast simulation is to solve the system for year t-1; i.e., a 1976 solution in early spring 1977. Broad features for the economy of t-1 are known. The largest countries have up-to-date reporting systems. Each separate model is adjusted, with add factors in some leading individual equations, to come as close as possible to correct or plausible values for t-1 (occasionally for t-2 as well). Unless there is firm evidence of a very temporary disturbance, the add factors are left at values set in t-1 and kept as "exogenous" input for t, t+1, t+2. Once a solution for t-1 is obtained, the dynamic sequence of solutions for the future years follows fairly quickly. At the time the forecast is first made for year t, some months of the year may have already elapsed, but very little is known on a quantitative level for the whole world economy at that time. As the year progresses and the solution continues to be refined, some preliminary indications of the year's outcome are accumulated, but they are not fully known until t+1.

The LINK system consists of some 5000 equations, and it is hard to summarize it in just a few variables. Records of LINK performance have been made for the past six years in terms of main world, country, or region magnitudes. These consist of world trade, world GDP growth, world inflation, and the same variables for individual countries or regions (Klein, 1976; Johnson and Klein, 1975). To summarize the main results of these retrospective error analyses, let me say

- The system has been perceptive in picking out the main movements in total world trade.

- Forecasts of volume have been better than forecasts of price movements.
- World and country GDP movements have been reasonably well forecast, especially some of the recession-recovery effects of 1974-1975.
- Individual trade flows have been estimated with fair accuracy, but estimates for some large regional groupings have been poor.
- The main imbalances, especially some that are persistent, have been indicated in advance.

Following a LINK forecast meeting during March, 1976, the central secretariat distributed in July, 1976, some post-LINK projections for 1974-1978. The historical years 1974-1975 were used for lining up LINK system relationships so as to reproduce, approximately, the main economic magnitudes. The system, so tuned, was then extrapolated to 1976-1978. It may be useful to summarize here the main results of the forecasting exercise, although the projection will be revised for the September, 1976, meetings of LINK.[1]

There is a very definite story to this forecast.

(i) World inflation is projected to decline gradually. This holds both for world export prices and world GNP (GDP) deflators. The former is a dollar denominated index and the latter in own-currency units. This has been a familiar LINK forecast position since 1974 and is basically working out in that way.

(ii) World trade, after falling in real terms (1974-1975)—a most unusual event—is expected to recover significantly and improve year-by-year over the forecast horizon. By 1978, world trade is projected to return to the vigorous growth path that it had enjoyed before 1973. This is a hopeful sign for the world economy.[2]

(iii) World production paused in 1975, with practically no change. There was a serious recession throughout the industrial (OECD) world. Its effect was felt in the developing world, and the Soviet grain failure contributed much to a poor showing in the socialist world. The poor Soviet performance was factored into LINK forecasts during summer 1975, partly as a result of discussion at the annual LINK meeting. The world recovery in production is strong but shows a tendency to slow a bit by 1978, subject, of course, to a generous allowance for forecast error.

This world summary, useful and indicative as it is, covers up a great deal of variability among countries or regional areas. Some output, inflation, and trade projections have been estimated for individual countries but are not tabulated here. Trade deficits for Italy and the United Kingdom persist in spite of currency devaluations. Also, France shows a tendency to

TABLE 1 Main World Aggregates

	1975	*1976*	*1977*	*1978*
World trade (bil. curr. $)	858.3	951.3	1075.5	1237.8
World export price (1970: 100)	2.144	2.282	2.416	2.504
(% Change)	(6.1)	(6.4)	(5.9)	(3.6)
World trade (bil. 1970$)	400.3	416.9	445.2	494.3
% Change)	(–5.0)	(4.1)	(6.8)	(11.0)
GNP (GDP) (% Change)	–0.22	5.51	5.52	5.17
Deflator GNP* (GDP) (% Change)	10.5	8.1	7.2	5.9

*Main OECD (LINK) countries only.

develop a trade deficit. The German and Japanese surpluses are noteworthy. The U.S. balance is in deficit in 1976, but the model fails to pick out the bad downturn that occurred in 1977.

A pattern of early strong growth in the United States, Japan, Germany for 1976 is supplemented by later strong performance in Canada, France, Italy, the Netherlands. The United Kingdom remains on a moderate growth path for the duration of this simulation. The developing countries gradually pick up towards their established growth targets after the slowdown in 1975-1976. The Soviet Union is also expected to perform closer to growth potential after the setback from the poor harvest of 1975.

The tendency towards improvement on the inflation front is seen to be pervasive across countries.

This example is a straightforward case of pure forecasting. A greater application potential is in the area of alternative simulations. Over the past several years, the LINK system has been used to study:

- exchange rate alternatives
- synchronized business cycle movements
- different (lower–towards 1973 values) oil prices
- alternative fiscal policies in the United States
- transfers to developing countries

Some of these have been reported in the second LINK volume (Waelbroeck, 1976; Johnson and Klein, 1974; Council of Economic Advisers, 1976).

POLICY USES AND IMPLICATIONS

Apart from the regular annual forecasts over future time horizons of three to ten years, we have attacked a number of interesting policy problems.

(i) Exchange Rates. Soon after the putting together of the LINK model, the leading nations convened in the Smithsonian (December, 1971) to put together an agreement that led to widespread exchange revaluation, floating rates, and a changed role for gold. The LINK system was used extensively to simulate alternative exchange rate policies and the implied international adjustments. These figured significantly in the international debate. The impact of exchange revaluations on the developing world were studied especially for UNCTAD. Exchange rates are assumed to be exogenous for each LINK simulation. They are not, however, constant and are frequently varied in accordance with policy assumptions and estimated trade imbalances. Alternative exchange rate assumptions for the D-Mark

and yen in LINK simulations for 1977 and 1978 were reported in testimony before the Joint Economic Committee Hearings, February 8, 1977. These exchange rate simulations were put forward in an unassuming manner purely for the analytic insight that they provided into the contemporary world economic outlook. In fact, the world press extracted these comments from the Committee hearings, and significant decline of the dollar on the Tokyo money market *was* clearly discernible within less than 24 hours. They also drew comment from Bundesbank officials. After this experience, LINK policy simulations were treated more confidentially.

(ii) Capital Transfers. On many occasions, LINK has made simulations for the UN on effects of transferring large sums of financial capital from the developed to the developing areas of the world, or from something like IMF gold sales to the developing world. These have been reported at various UNCTAD meetings and other North-South forums. Generally speaking, the capital transfer simulations showed that developing countries were enabled to import to greater advantage as a result of the transfer and that industrial countries would improve their trade position by having better export markets in the developing countries. World trade and activity as a whole also benefited in these simulations. The total sums considered were usually about $20 billion.

(iii) Oil Embargo. The actual embargo, subsequent OPEC pricing of oil, and world recession were simulated and discussed in professional or semipopular conferences on contemporary international economic conditions. In this respect, special tabulations were made for the U.S. Treasury and other U.S. government agencies to try to assess the impact of the turbulent developments in the world economy. Many LINK simulation results for these problems were reported in some detail in the world press. A lengthy account, right after the onset of the crisis of autumn 1973, was reported in the New York *Times* (January 27, 1974). A major conclusion of these calculations was that the synchronized shock imposed on the world's oil importing countries would have mutual amplification effects that would generate a world recession on the order of the 1957-1958 world recession. When these results were reported to a meeting at the U.S. Treasury in December, 1973, it was urged that countercyclical measures be introduced to deal with the impending recession.

(iv) World Recovery. In testimony before the U.S. Congress, in professional meetings, and in general distributions to the media, policies of economic stimulus by certain countries with selected exchange rate changes have been analyzed through the LINK system. These results have been widely distributed through U.S. government departments and international agencies. When Helmut Schmidt was urging the United States to reflate, and thereby help the whole world recovery, in contrast to his

stances of 1977 when Germany was being asked to do the same, the LINK Project was called upon to estimate the world wide consequences of the United States' taking stimulative action that would have *increased* our growth rate by one full percentage point. The general conclusion was that only a negligible improvement in the world growth rate would have resulted from the addition of one point to the American rate in 1976. This is not to belittle the stimulative value of the much larger swing in U.S. growth rates between 1975 and 1976, only the marginal contribution of one more percentage point. This calculation was reported in the *Economic Report of the Council of Economic Advisers to the President,* January, 1976.

Similar studies on stimulative policies by Japan, Germany, and the United States were made in 1977 for U.S. government offices but reported only in internal memoranda.

(v) Commodity Stabilization. For UNCTAD, the World Bank, and other world agencies, LINK has examined the question of raw material pricing on world markets, the effects on different economies, and the possibility of initiating price stabilizing buffer stock schemes. LINK has actually constructed commodity market models and integrated them into the whole system for interested international agencies. It is a matter of the project's pursuing an interesting research area and at the same time trying to contribute to international discussion of an important policy issue. At a conference at Airlie House, Virginia, March 1977, F. G. Adams presented results in COMLINK, a merged simulation between world primary commodity models and the LINK model. It showed that sagging world commodity prices in 1977, estimated in detail from satellite commodity market models, reduced estimated activity for primary producing countries and lowered the LINK estimate of major world aggregrates. Many policy-making economists from United States and international agencies participated in this conference.

These are the main lines along which LINK has participated in the policy process. New research work is under way on improving the long term capability of LINK simulation in the hopes that the Project will be able to contribute more to world discussion of long-run growth potential. This research is strongly supported by the United Nations, where long-run economic development is an issue of major concern. The Project started from a concern with the short-run business cycle transmission mechanism and stabilization. It is natural to try to extend it to cover growth. A new forecast policy round of simulation is initiated at the beginning of every year. The results of these calculations are discussed in worldwide LINK meetings each March and September, with revised calculations being made on the basis of country-by-country information discussed at the meetings.

At year end, a general press release giving forecasts of broad tables is provided to the popular media. At the summer meetings in Japan in 1977, the outcome proved to be sufficiently pessimistic that we agreed on a world countercyclical policy statement presented to the Japanese and foreign press at the close of the meetings.

A second area of research is in capital flows and exchange rate determination. LINK has already had an impact on exchange rate policy, but could have a more decisive influence in the future after capital flow modeling has been strengthened in the system. Floating rates are here to stay, for some time; therefore, LINK attention has turned towards endogenous generation of rates. The same things could be said of commodity market modeling. That area of research will be extended and improved. When it is further along, it should be able to make a greater impact, particularly in North-South debates.

CONCLUSION

Since the Project's inception, the computer programming for LINK has improved. It is now more easily transferred and is available on several computer installations in the United States, Europe, and Japan. Eventually, it will be on a worldwide network, available for policy application by more international users. There is even the possibility of simultaneous interactive use internationally, simulating policy moves from country to country.

I would like to see Project LINK eventually turned over to a major organization for its regular usage and work on further development. I have in mind an institution like an international development agency which would maintain it and employ it for regular forecasts and simulations. While the intellectual *independence* of an academic location for central LINK activities is much to be desired, the work load has grown, and continues to grow, to such an extent that larger research and analysis facilities are needed than can usually be obtained in an academic environment.

In the formative stage when freedom for creative thinking was necessary and the project was small, the academic environment was ideal for the central facility. After 12 years of development work, however, the project has grown to the point where many data clerks, programmers, secretaries, and general research assistants are needed (all these are in addition to senior researchers). It is not unusual for a research tool to be developed in academia and then turned over for full implementation by an official agency.

One department of a national government could readily handle the details—Federal Reserve Board, Treasury, State, Commerce are all plausible U.S. locations. Foreign participants, who are essential to the very concept of LINK, would probably not function well with the project located in a U.S. national agency. Accordingly, a supranational agency is needed for performing all the necessary host functions. Such an institution could be:

- International Monetary Fund
- World Bank
- United Nations
- Bank for International Settlements
- OECD
- GATT
- Common Market
- International Institute of Applied Systems Analysis

Some of these are more natural than others for the task, but if primed, any one could undoubtedly rise to the occasion. The project could function much as it does now, but all the secretarial work, numerical analysis, basic research forecasting, and so forth could best be handled through the research department of one (or more) of these international institutions.

TECHNICAL APPENDIX

The formulas for consistent international linkage are based on the concept of the world trade matrix, which serves as a centerpiece of the whole system. This matrix shows the dollar value, in current or constant prices, of the bilateral flows between countries or regions. It is presently being prepared annually by the LINK project with about two years' delay. It is, by itself, a highly informative tabulation of the structure of world trade. It is available by commodity classification.

A typical element of this trade matrix will be denoted as: X_{ij}—the flow of trade from country/region i to country/region j. A matrix of coefficients is formed:

$$a_{ij} = \frac{X_{ij}}{M_j}$$

showing the ratio of j's imports from i to j's total imports. Denote the coefficient matrix by A. We then have the matrix equation

$$E = AM \qquad [1]$$

This shows that imports, computed from individual models (in the column vector M) are transformed into exports (in the column vector E). An element of E is therefore a weighted sum of partner country imports, the weights being the corresponding row of A. Since the column sums of A are unity, by construction, it follows that world trade is the same whether measured as the sum of E or the sum of M elements.

Let us assume that E and M are measured in constant dollars. (An equivalent analysis could be worked out for the current dollar case). The world trade identity must hold also in nominal terms.

$$(PE)'E = (PM)'M \qquad [2]$$

(PE)′ and (PM)′ are row vectors of export and import prices.

By substitution of equation 1 into 2, we obtain

$$(PE)'AM = (PM)'M \qquad [3]$$

This is an identity in M. Equating left and right hand sides, term-by-term, we find

$$(PM)_j = \sum_{i=1}^{n} a_{ij}\,(PE)_i \qquad j = 1, 2 \ldots n. \qquad [4]$$

This shows how import prices can be obtained as a column weighted sum of partner countries' export prices.

These are four basic LINK equations that are used to tie the different models of the system together.

THE LINK ALGORITHM

A standard procedure is used to solve the LINK system so that the world restrictions on equality between exports and imports are preserved. If the specialists of each country are left to their own information for making economic projections from solutions of their particular models, there is no mechanism for insuring that the total of exports will equal the total of imports. In fact, the usual situation is that countries/regions, as a whole, will expect to export more goods in the world economy than the same countries/regions will expect to import. It is not necessarily the case that exports so determined will exceed imports, but it usually works out that way.

To start a solution (projection or simulation) of the LINK model for the world economy, each individual model is solved by itself in pre-LINK mode; i.e., for given assumptions about the country's or region's exports and import prices. In a symbolic sense, we can express any endogenous variable of a model as a function; in reduced form, of exogenous inputs including exports and import prices.

$$M_i = f_i(E_i, PM_i, \ldots) \qquad [5]$$

$$PE_i = g_i(E_i, PM_i, \ldots) \qquad [6]$$

The pre-LINK solution of the model for country i will be determined on the basis of assumptions about exports and import prices, together with assumptions about other input values associated with domestic policies and other exogenous variables. These import values and export prices so determined from f_i and g_i for each country/region are then substituted into equations 1 and 4. The pre-LINK values are in own-currency units and must be converted at exchange rates into U.S. dollars for use in equations 1 and 4. After values for E and PM are computed from these trade matrix equations, they are converted into own-currency units before being put back into individual models. It is generally the case that the values of E_i and PM_i will be different from those used in the original pre-LINK solutions.

This set of calculations will be iterated until the values of E_i and PM_i do not change from iteration to iteration. Actually the convergence criterion is implemented on iterated values of $\sum_{i=1}^{n} E_i$ (total world trade). The equality between world imports and world exports is guaranteed by the structure of equations 1 and 4. This is a general statement of the algorithm.

There is one major complication in the LINK methodology. It is implicitly assumed that the trade matrix A is constant. Over historical periods, with some time delay, there is an annual observation of A_t, and we could use a different A_t for each year's solution, but in forecasting, it is not possible to know A_t in advance. Equation 1 is an identity if written as

$$E_t \equiv A_t M_t$$

On the other hand, if a base period value A_0 is used, the equation must be

$$E_t = A_0 M_t + R_t$$

where R_t is a "residual" vector. There are two options for research:

(1) Model the elements of A_t in order to project the whole trade matrix through time or
(2) Model R_t to project the movements of total country exports through time. Both methods are being investigated within the LINK project.

It is easier, in a sense, to model R_t because there are fewer elements in R_t than in A_t. An obvious specification is to use principles of the linear expenditure system (LES). It has the property of preserving budget restrictions in consumer theory, and these can be immediately adapted to the preservation of the accounting identity between aggregate imports and aggregate exports. The LES also has the property of taking account, in a simplified way, of relative price effects on demand. In the present case, the analogous idea is to take account of relative changes in international prices to explain changes in trade. In particular, the changes that are associated with exchange revaluations, energy prices, food prices, and other raw material prices are important in the context of international trade analysis of the recent and present period.

Country i's exports can be computed from

$$(PE)_{it}E_{it} = \alpha_i(PE)_{it} + \beta_i(A_{oi}(M\$)_t) - \gamma_i \sum_j \alpha_j(PE)_{jt} \quad [7]$$

In this equation, A_{oi} is the ith row of the base period (o) trade shares matrix A_o, where A_o is understood to be evaluated from current-priced trade flows. $(M\$)_t$ is a column vector of imports, evaluated in current US\$. If $\beta_i = 1$ and $\sum_i \gamma_i = 1$, we have the full restrictions of the LES and the adding-up property holds

$$\sum_i (PE)_{it}E_{it} = \sum_i A_{oi}(M\$)_t$$

or

$$\sum_i (E\$)_{it} = \sum_i (M\$)_{it}.$$

The export equations are written in static form, but time trends for taste or technical changes and lags may be introduced in order to make the export side dynamic. In practice, we have usually estimated β_i by regression methods (slightly above or below unity) and γ_i (also by regression) by assuming an a priori weighting scheme for $\sum_i a_j(PE)_{jt}$. The results show, empirically, that the world identity holds within \$1-\$2 billion in a total of \$800-\$1000 billion. In some special calculations, however, we have also estimated the LES system under full constraint. These constrained equations are presently being tested in the LINK system.

Alternative approaches based on estimation of elasticities of substitution or similar concepts have been reviewed (Gana et al., 1979; Amano et al., 1980). Eventually, we may hope to be able to develop estimates for all bilateral trade flows separately, as in the Federal Reserve System's multicountry model (Berner et al., 1977). Their model is small, consisting of only five countries, but data and methods are being developed to do this on a larger scale (Ranuzzi, 1979).

NOTES

1. The division of the change in world trade at current prices into a volume component and price component was changed so that volume was projected to grow relatively more strongly in 1976. Also, the strong rate of growth for 1978 in world trade was scaled down slightly, while 1976 and 1977 were raised.

2. In fact, world trade increased by a large amount (11%) in 1976, and by a moderate amount in 1977 (5%). The forecast for 1978 is for another moderate increase; therefore, the yearly pattern of growth in Table 1 has been reversed, 1976-1978.

REFERENCES

AMANO, A., E. KURIHARA, and L. SAMUELSON (1980) Trade Linkage Submodel in the EPA World Economic Model. Economic Bulletin 19. Tokyo: Economic Planning Agency.

BERNER, R., P. CLARK, E. HERNANDEZ-CATA, H. HOWE, S. KWACK, and G. STEVENS (1977) A Multicountry Model of the International Influence on the U.S. Economy. International Finance Discussion Paper 115. Washington, DC: Federal Reserve System.

Council of Economic Advisers (1976) Report to the President. Washington, DC: Government Printing Office: 135.

GANA, J., B. G. HICKMAN, L. J. LAU, and L. R. JACOBSEN (1979) "Alternative approaches to linkage in national econometric models," in J. Sawyer (ed.)

Modelling the International Transmission Mechanism. New York: Elsevier North-Holland.

JOHNSON, K. and L. R. KLEIN (1975) Error Analysis of the LINK Model. LINK Working Paper 7. Philadelphia: Economic Research Unit, LINK Central, University of Pennsylvania.

--- (1974) "LINK model simulations of international trade: an evaluation of the effects of currency realignment. "J. of Finance 29 (May): 617-630.

KLEIN, L. R. (1976) "Five-year experience of linking national econometric models and of forecasting international trade," in H. Glejser (ed.) Quantitative Studies of International Economic Relations. New York: Elsevier North-Holland.

RANUZZI, P. (1979) Euro-LINK Model: Bilateral Trade Linkage Equations. Brussels: European Economic Commission.

WAELBROECK, J. [ed.] (1976) The Models of LINK. New York: Elsevier North-Holland.

6

GLOBAL MODELING AND DECISION-MAKING

Mihajlo D. Mesarovic and Barry B. Hughes

Global modeling has received wide recognition in conjunction with the activities of the Club of Rome. To begin with, some observations are in order on the motivation for an effort to construct a global model which can be used for practical decision-making. There are three new dimensions in international relations today:

First, issues which have traditionally been considered separate are becoming more and more interrelated. Not "first we have to do this" but "first we have to do everything." (Cleveland, 1976). There is no way to formulate a comprehensive energy policy without relating it to international trade and balance of payments, which cannot be formulated without a food and agricultural policy, which relates to the consumer price index and inflation, which influences unemployment, production, and energy policy. There is a circular relationship which requires all elements to be considered simultaneously.

Second, there is the interdependence between nations. Unemployment in Detroit is affected by the world price of oil, while inflation in the price of food in the United States might depend on the weather and climate variations being experienced by Indiá and the USSR. The United States is importing at the present more than 40 percent of its oil, 80 percent of its nonferrous primary materials, and close to 70 percent of all ferrous materials excluding iron. A world economic system is developing in which interdependence among nations is increasing. A new pluralistic world system is emerging "with nobody in charge."

Third, continuity of progress of the world as a whole is being questioned. For a variety of reasons, from physical and natural resources constraints to organizational management, social and psychological, it is being questioned whether progress with an ever-increasing standard of living, defined in whatever way one chooses, can continue uninterrupted, whether there are "limits to growth," and what are the alternatives to the growth ethic.

These issues are both important and complex and must be considered in an explicit and integrated manner. Systems analysis techniques provide a basis for conducting such an investigation.

SYSTEMS ANALYSIS TECHNIQUES

The hallmark of the systems analysis techniques used in this effort are the following: All the interrelated issues are given the same attention; development of physical resources, demography, economic development, industrialization, and so on, are given equal attention and are represented at comparable levels of detail. Whether one is concerned with national or regional problems, one can use systems analysis techniques to treat these problems in the global context, which is increasingly becoming a necessity for any practical and realistic consideration of alternatives. Furthermore, the policy alternatives can be considered in a long-term framework in which the question of direction, magnitude and the meaning of growth is explicitly included.

Because of the importance of these new directions in international relations, global modeling as such has been taken very seriously by responsible decision makers in many different parts of the globe. A variety of national and international organizations, from national governments to the UN and the OECD, and other forums have initiated concentrated efforts at formulating their own approaches to global modeling with the use of a variety of systems analysis techniques. It could therefore be said without hesitation that global modeling as an activity has come of age in the sense that it has been taken quite seriously by a broad spectrum of professional and political groups. It is to be expected that it will become an even more important activity over the next five to ten years; it might be argued that almost any professional international group will have to pay attention to the global aspect of its activity and will need to be mindful of the results of global modeling efforts.

MULTILEVEL REGIONALIZED WORLD MODELING

The second comment we would like to make refers to a specific global modeling effort in which we have been involved over the last ten years. In

1971, after the MIT group had proposed the use of the systems dynamics approach on the global modeling requirements of the Club of Rome, Dr. Eduard Pestel from the Technical University of Hannover decided to search for alternative systems analysis techniques which, in his opinion, could provide more richness to the exercise and also come closer to meeting decision-making needs. He approached us at that time with a request to formulate jointly a long-term project based on multilevel hierarchical systems technique to the analysis of complex issues and on general systems theory. After hesitating to jump into the cold water for a while, we initiated what has come to be known as multilevel regionalized world modeling activity.

The first phase of that activity was centered on preparation of a report to the Club of Rome concerned with strategies which could be designed globally to avoid major mishaps, if not catastrophes, stemming from the limitation of resources and lack of rationality and planning in using them. Implicitly, the notion of the limits to growth has been accepted and it is assumed, at least as far as use of resources is concerned, that mankind is moving toward an equilibrium condition. The question with which the report was concerned was how such equilibrium could be reached at a sufficiently high level of prosperity and without reversals.

In our effort, we were concerned with the time dimension of the problem and with the urgency of taking some actions; therefore, we had to be reasonably specific as to where decisions and actions had to be taken. In view of this, instead of considering the world system as an integrated whole (as was the case in the earlier Club of Rome efforts), we have represented the world system as the set of interconnected subsystems and have recognized that one of the key problems to today's world is that while the world is becoming one—even more strongly interdependent—it is also becoming many, i.e., the gaps between the increasingly interdependent parts of the world are increasing too. Two conclusions from that analysis have been specifically identified: First, we have concluded that by anticipatory thinking and global use of resources, major reversals impeding mankind's development *could* be avoided; that is, the limits to growth in reality need not prevent us from achieving an equilibrium which is at a sufficiently high level. Second, to achieve such equilibrium, the concept of organic growth has been proposed. In essence, organic growth means selective growth in different parts of the world with increased functional interdependence so that the world system would operate increasingly as an "organism"; that is, as a system where, if any of the subsystems were to be isolated, it could operate only at a lower level of satisfaction.

Although we tried in the report (*Mankind At The Turning Point*) to be very optimistic, it was greeted with mixed emotions and again was considered as a doomsday prophecy. As a matter of fact, some critics have considered the report to be even more pessimistic than the famous *Limits to Growth* study. The reason indicated for such an assessment was that while it is true that the report identifies possibilities at the level of physical resources which could avoid reversals in the progress of mankind, conditions needed to achieve that level would apparently require social and international actions which appear to be unrealistic and outside the domain of practicing politicians' serious consideration. The second remark made about the report was that it was prepared by people who had their roots in a Western culture and therefore could not claim to have an unbiased view as to the future desires, aims, and development strategies in other parts of the world with different social and cultural conditions.

In response to these criticisms, we have initiated the second phase of our modeling effort which has the following characteristics: First, we tried to bring the concern from the general global or regional level to the national level; that is, to the level where actual decisions could be made. Even in the regional study, we did not assume the existence of either regional or global authority. Yet, the strategies were couched in regional language. It was necessary, therefore, to bring the focus of consideration to the national level. Second, we had the objective of addressing the issues which decision makers in different parts of the world are facing in practice so that recommendations for action could be dealt with in real terms. Third, we had the intention of bringing into the group concerned with the long-range global assessment people from other cultures to complement and improve our views of the global development process. The second generation World Integrated Model (WIM), in which we adopted this approach, is outlined very generally in the Technical Appendix.

In order to achieve these objectives, we have done several things: First, we have constructed a long-range planning and policy assessment tool, termed the APT-system (Assessment of Policies Tool), which can be used for the assessment of long-range development by anyone who will formulate a scenario. A scenario is defined as a sequence of assumed policies and a set of events concurrent with these policies. Second, within this planning and policy assessment tool we have implemented an improved version of the world system model in which a nation is shown explicitly so that it can be used by the national decision makers. Description of the model is given in the Technical Appendix. Third, we have initiated the APT program aimed at bringing the APT-system to practical use at decision-making levels.

APT PROGRAM

The APT program involves a set of projects, fully and completely independent from each other in aims, personnel, organizations, and financial support. The common bonds of all the projects consist in the general objective of implementing an appropriately constructed APT type system for practical and *immediate* use in specific contexts; and reliance on the common conceptual and methodological heritage of the world modeling project which led to the publication of the Second Report to the Club of Rome.[1]

In addition to fulfilling its aim of bringing a global policy assessment tool closer to the national and international political decision-making domain, the APT program intends also to organize demonstrations of the use of the global planning tools to selected groups of interested policy makers and the public at large in different parts of the world. The projects under way or being developed in different parts of the world are listed in Table 1.

What has emerged as the organizational format for the APT program is an informal association, in which the bond between different teams is based on being concerned with the same real life issue but translated and interpreted in a specific context relative to geography and problem focus. For example, the Iranian group was concerned with the problem of food self-sufficiency and the need to decide which of the exporting sectors should be given priority. The German group was concerned with national R&D policies and their impact on European and global development. Other groups, which are joining in now from Indonesia, for example, will have their own specific concerns and yet will have a bond which links them with the informal association of the APT projects. Each one of these problem concerns is a specific and national one which, at the same time, can be neither considered nor solved outside the long-term and global context. How to put a national issue into the global context is what brought the various groups together. The APT project is an example of a group which is based on neither professional affiliation nor government representation. Rather, it is based on an exchange of views between people who have an urgent problem to solve, who do not otherwise possess a satisfactory method for devising a solution, and who feel that by exchanging information they can get information in return which will help them meet an otherwise impossible challenge. Since each and every one of these groups remains within a given decision-making structure (as it prevails in a given nation, region or international organization), their concern and their output must and does remain at a practical and applicable level. Therefore, if successful, by definition the group will have an impact on decision- and policy-making. The absence of any formal organizational link allows them to remain fully within the organizational structure as it exists in the given national, regional, or international setting so that they do not have to compromise their position in the process of exchange. The question of confidentiality, of proprietary types of information, and so on, is

TABLE 1 Partial List of Projects—Active or Under Development—in Which WIM World System Model and APT Planning Tool is Used for Practical Policy Analysis

1. "Impact of Research and Development Policies in Federal Republic of Germany on the Long Term National Economic Development"—*Federal Ministry of Research and Technology, Federal Republic of Germany.*
2. "Impact of the Alternative Global Development on the Long Term Prospects for Iranian Economy"—*Planning and Budget Organization, Tehran, Iran.*
3. "Integration of the WIM World System Model, GOL World Agriculture Model and NIRAP U.S. Food Model for the Long Term Assessment of US and Global Food Policies"—*U.S. Department of Agriculture, Washington, D.C.*
4. Project LOPROLA—"Alternative Future for Latin America"—*Fundacion Javier Barros Sierra, Mexico City, Mexico.*
5. "Long Term Energy Policy Options for U.S."—*Edison Electric Institute, New York, U.S.A.*
6. "Pacific Region Year 2000"—*East West Center, Honolulu, Hawaii; A multinational project with participants from Australia, Indonesia, Togo, Japan, Philippines and U.S.*
7. "Scenario Analysis of the Prospects for North/South Relationships"—*Ministry of Foreign Affairs and President's Office, Republic of France, Paris, France.*
8. "Consistency of Alternative U.N. Targets for Long Term World Development—*Center for Development, UNITAR and UNIDO, United Nations, New York.*
9. "Project ECOMIN—Analysis of Man-Environment Interaction on a Global Basis"—*A joint USSR-USA project under development.*
10. "Analysis of Potential for West African Common Market"—*A project being considered jointly by West African countries and European Common Market.*

handled exactly the same as in any other contacts and procedures; no special government approval either has to be asked or granted for such an exchange of information. Again, what keeps the group functioning is that there is an outstanding problem (that is, how to deal with long-term decision-making in the global context) which some of the groups separately can solve in this reality which brings and keeps the groups together.

SIMULATING ALTERNATIVE FUTURES OF THE WORLD SYSTEM

There is an event being planned which, in a sense, will present the highlights of the APT program as it has developed to date. It is anticipated that an experiment will be conducted of a joint and interactive scenario analysis of global development. In that event, computer consoles will be placed in different parts of the world—from Jakarta and Tokyo to Caracas and Tehran—all connected with a data bank and a world system model in the central location. At each of the consoles there will be an expert group fully acquainted with the functioning of the model, with the methodology of formulating scenarios, and—most importantly—with the goals and desires, plans and strategies (preferable and available) for the given nation or region. Each one of the groups will formulate a development plan for a fixed period of time (for example, five years) which will be transmitted via satellite communication and entered into the central computer system model. The computer will respond by describing at the end of the planning period the state of the planet (e.g., as of five years hence) which information will be transmitted back to the participants. By reviewing the results of the policies and the newly emerged state of affairs, new plans will be formulated and implemented and the state of the world system after yet another increment of time is then assessed. By doing such an incremental analysis, a plausible scenario for the world's future up to the year 2000 and beyond will be prepared.

The importance of this exercise stems from the fact that each of the participating groups will use the same agreed-upon data bank and the same relationships to represent the world development system. Therefore, whatever outcome the scenarios and model-use produces, it will have credibility inasmuch as it will reflect differential cultures and views of the future incorporated in its use in a more realistic fashion than in any known previous attempts. In addition, an opportunity will be given to the participants to repeat the event several times and, in that process, to be confronted with some unexpected events; for example, weather or climate reversal, decreased or increased population growth, a nuclear power plant accident, faster or slower depletion of resources, and so forth. In such a way, several alternative futures will be designed. It is intended that the results of this joint worldwide analysis will be presented to a high-level international group of policy and decision makers in order for them to assess the outcome as well as to consider establishing a network of information and knowledge exchange between countries in different parts of the world. Those nations which are concerned with the long-term global future can then utilize this data and knowledge in practical decision-making. If such a network is indeed developed, it will represent a practical step toward global rationality; it will represent a "white line of knowledge"; a positive link rather than a defensive one, such as the famous "hot line" between the superpowers represents.

This event may very likely represent the culmination of the APT program and possibly its end. A new structure, a new set of relationships, will then have to be designed. An important lesson from this may be derived: In order to remain practical—to stay close to reality and influence decision-making in a given organizational context—the global or international group will have to be problem-oriented, fully observe national and regional constraints, and be prepared for its own dissolution; neither to go out with a bang nor with a whimper, but simply to be made obsolete by its own success. It is this flexibility and dynamic which in real life is the essence and inevitable outcome of success. There is nothing more damaging for any movement or group than achieving final success; and this dictum, we believe, will have to be accepted as a fact of life and as a requisite for avoiding sterility of many international organizations, be they professional or intergovernmental, even the United Nations.

TECHNICAL APPENDIX

Block Diagrams of Selected World Integrated Model Structures

The second generation world model, referred to as WIM (The World Integrated Model) represents the global system in terms of 12 geopolitical and economic regions, and has the capability of identifying up to 5 nations or subgroupings within each of the world regions. It therefore allows consideration of long-term national plans in regional and global context.

The world model consists of several submodels, specifically population, labor and education, aid and loans, economics, food and agriculture, energy, materials, investment goods, and trade submodels. This description is only illustrative of the model and will focus on the energy submodels. Other submodels exist at similar levels of detail and with comparable structures. These other models are described elsewhere (Hughes, 1980). Flow charts will be used to describe the structure of the submodel.

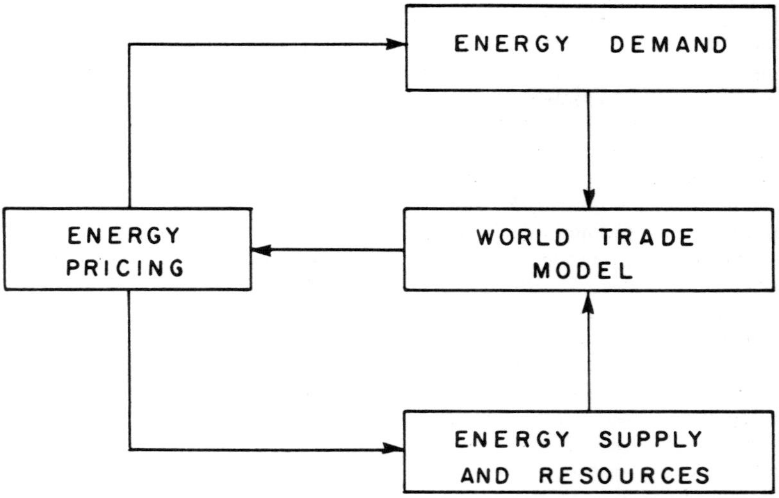

Figure 1: Major Components of the Energy Submodel

It is useful to mention the overall structure of flow for the computations in the world model. The computations begin in the population and economic models by establishing preliminary estimates of population and economic growth and distributions. The flow of the program next moves to the physical models, in which production or supply capabilities and demand are calculated. The program flow then moves to the world trade mechanism, in which supply and demand on both regional and global bases are balanced and in which balance of payments situations influence total exports and imports and also influence the pattern of imports for regions with balance of payments problems. After the determination of trade patterns and final demand and supply in the physical commodities, the model flow returns to the physical models. At this point, actual production and demand are recorded and investment decisions and prices are calculated on the basis of surplus capacity and anticipated demand patterns. Thereafter, flow again moves to population and economic models in which the investment decisions and food availability coming from the physical models have their impact. This ends a given time cycle of the model and a new cycle begins.

Let us begin to look at the flow charts and structure of the various energy submodel components. Figure 1 shows the basic elements in the structure of the energy model. At the top of the figure, energy demand is determined on the basis of values coming from the economic and population submodels. It is also, of course, affected by the pricing of energy. At the bottom of Figure 1, energy supply is a function also of the economic model (specifically investment availability), of energy pricing, and of course of the energy resource picture. Energy demand and energy supply on a regional and on a global basis are reconciled in the world trade submodel. The reconciliation of energy demand and energy supply in that model produce information about surplus capacity, which has a major impact on energy pricing and other components of the energy model. Energy pricing, as we have noted already, affects demand, supply, and, of course, extraction of resources. Let us proceed now to look in more detail at the basic components of this energy structure. First, at energy demand, next at energy supply, and finally at energy trade and pricing. It should be noted, of course, that all other submodels, specifically food and agriculture, materials, investment goods, also have important implications through the dynamics of the model for the energy model.

Figure 2 is a flow chart showing the principal variables in the energy demand subunit. At the upper lefthand corner are variables which determine the total energy demand. Specifically, total energy demand is produced by an energy demand function in which the principal elements are gross national product, gross national product per capita, energy demand elasticity, and, of course, the price of energy. The specific details of this energy demand function have been described elsewhere but we can touch briefly on that structure. Energy demand is a function of the gross national product and energy demand per unit of the gross national product. Energy demand per unit of the gross national product is in turn a function of the level of economic development of a region represented by the gross national product per capita; that is, the income elasticity of demand is a function of economic development. The basic energy demand per unit of gross national product is, of course, influenced by the price of energy and the elasticity of demand with the price of energy. Once total energy demand for the region is specified, we move to the problem of specifying energy demand for the various energy types within the model structure. At the current time the model specifies demand, supply, trade and consumption in five energy types: oil, natural gas, coal, hydroelectric, and nuclear power. The basic mechanism for dividing energy demand into energy demand for the various primary energy types is through the use of the demand vector. The demand vector represents basic consumer preferences for energy types, and it is modified or adjusted by the price of the various energy types and cross-demand elasticities. The result of the energy type demand function then becomes specific energy demand for the various primary fuels in the energy model. These are represented at the bottom righthand side of the flow chart as energy demand for type 1 through energy demand for type n. We will see that these energy demands are taken to the world trade model and final energy consumption is computed there.

Figure 3 shows the flow for the energy supply subunit. The ultimate product of the energy supply subunit is the production capacity for energy types 1 through n shown near the bottom of Figure 3. This is then parallel to the structure for energy demand from which the ultimate products were energy demand for all energy types. Production capacity, like energy demand, is taken to the world trade subunit for reconciliation and balance. Production capacity is a function of the capital stock for the energy type and the capital energy output ratio. Most of our discussion of the energy supply subunit will be directed toward explaining how energy investment decisions are made and thus energy capital is incremented. Not shown on the flow charts is the decrementing of energy capital with depreciation determined by the lifetime of capital in the specific energy type.

Energy investment, as shown in the bottom lefthand corner of Figure 3, is produced as an output of the energy investment function. The primary inputs in the energy investment function are the total investment available in the economy, the price changes in the energy type and the elasticity of investment with those price changes, and the reserves of the energy type which serve as a constraint on the investment pattern. More specifically, the initial or basic share of investment from the economy going to a specific energy type is changed as the energy price changes through the mechanism of the investment elasticity. It has been found that this

(text continues on page 120)

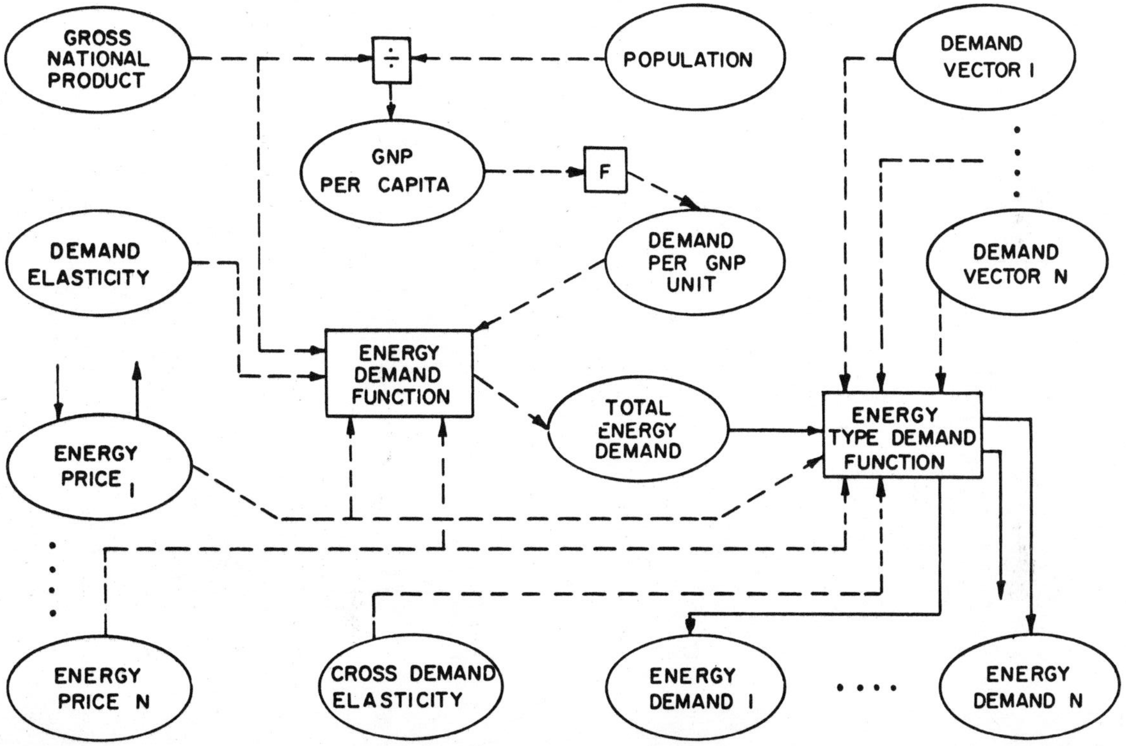

Figure 2: Energy Demand

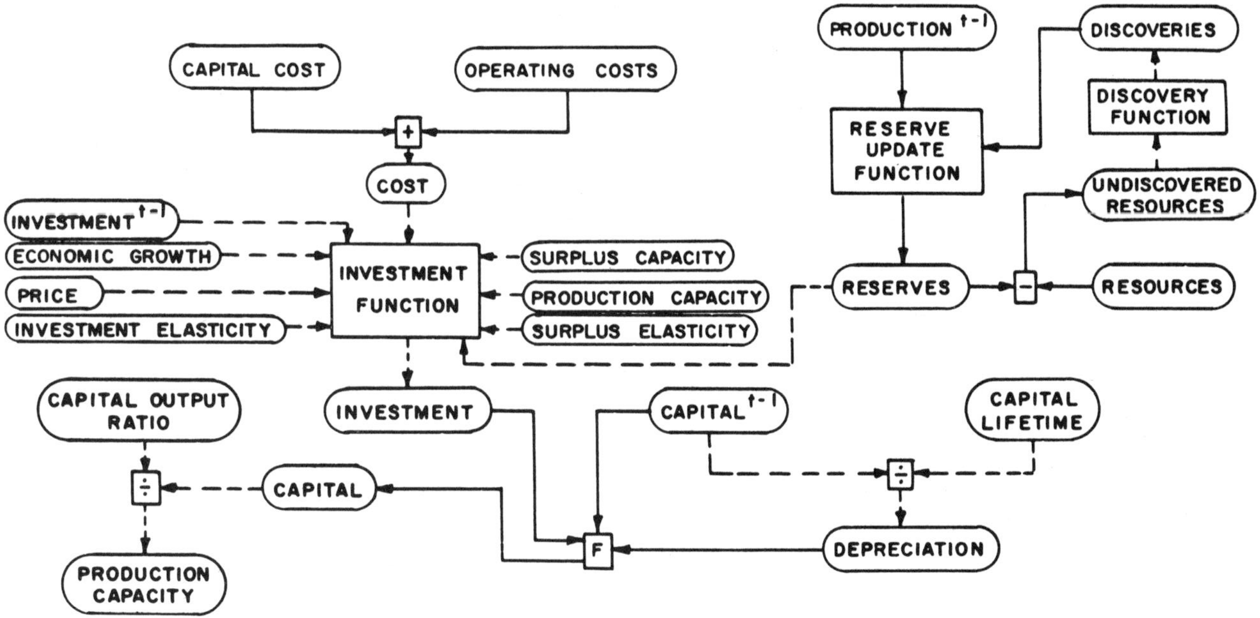

Figure 3: Energy Supply

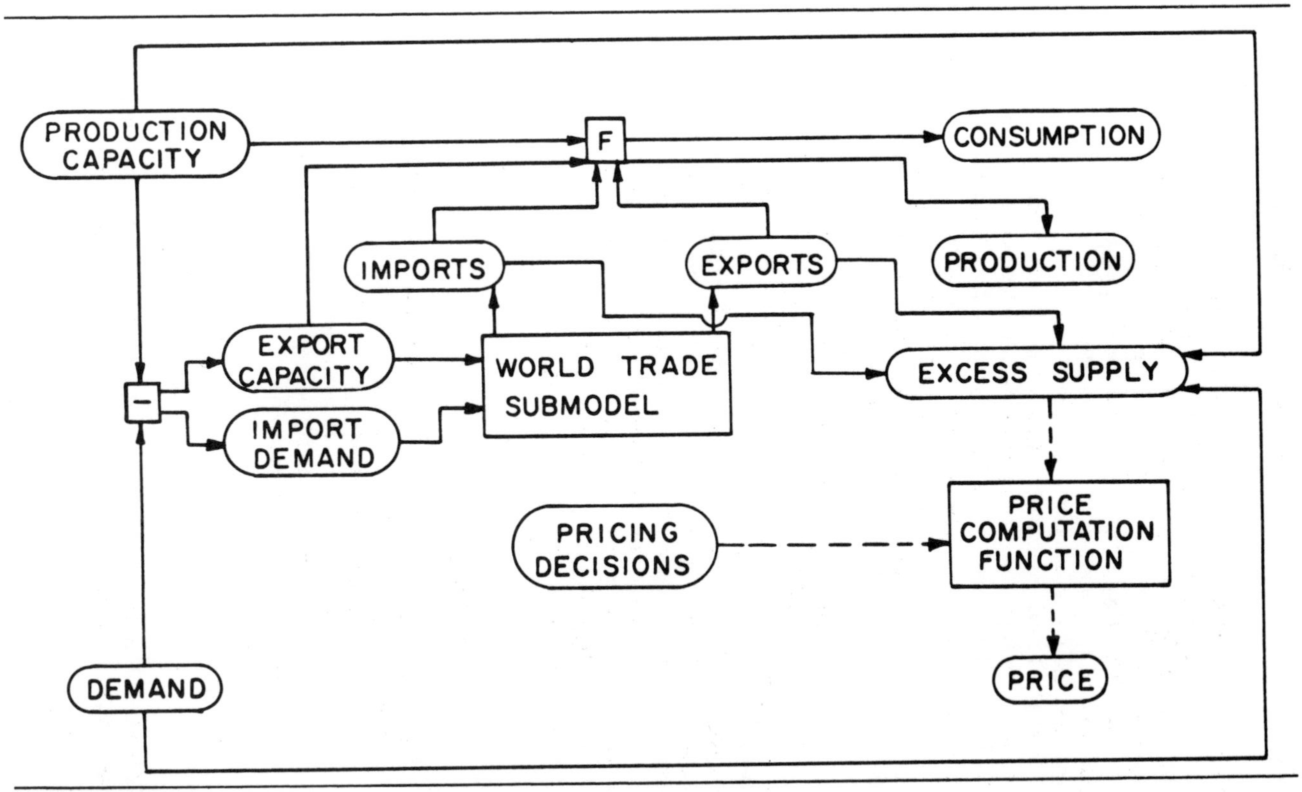

Figure 4: Energy Trade

mechanism, whereby investment is changed with price rather than supply being changed directly with price, more adequately represents the delays in supply change with price than does a simple supply elasticity function. Moreover, in a period of price changes, the supply elasticity must be represented as short-term and long-term and, in fact, should be represented as a completely graduated supply elasticity with time. Since this is normally impossible or extremely difficult, the use of an investment elasticity is much more appropriate and functional.

Naturally, energy investments are not made when the lifetime of reserves either for energy types or other raw materials drops too low. We represent this fact by a constraint on investment coming from reserves. Reserves are calculated on the basis of discoveries which increment reserves and production which decrements reserves in the reserve update function. Discoveries, in turn, are a function of the undiscovered resources represented in the model; the undiscovered resources, of course, being the difference between ultimate discoverable, and at the current price recoverable, resources and reserves.

Now let us turn our attention to Figure 4, which shows what happened to the model after we obtained the energy demand and the production capacity for each of our energy types. Figure 4 illustrates what happens for one of the energy types. Knowing demand and knowing production capacity, we can compute import demand and export capacity on a region-by-region basis. These are primary inputs into the World Trade Model. Coming from the World Trade Model are actual exports, final consumption, actual imports, and information concerning surplus capacity for unfulfilled demand on an energy-type by energy-type basis. The information concerning surplus demand for deficit capacity goes into a price computation function, which allows either exogenous pricing decisions or energy pricing on the basis of supply-demand targeted at the maintenance of a small surplus capacity at all times.

NOTE

1. A theoretical description of an early version of that model is available in the Proceedings of the IIASA (International Institute for Applied Systems Analysis), Laxenburg, Austria, and can be obtained upon request. An updated version of that material will be published as a series of monographs by North-Holland, Amsterdam. A computer listing of the program and the tape containing that program is available at the cost of photocopying or reproduction.

REFERENCES

CLEVELAND, H. (1976) The Third Try at World Order. New York: Aspen Institute.
HUGHES, B. B. (1980) World Modeling. Lexington, MA: D. C. Heath.

COMMENTARY

Robert W. Russell

Harry Johnson's discussion of two international "networks" of professional economists, the Bellagio group and the Mundell groups, is a veiled critique of the naive model which assumes that international scientific and professional associations can persuade policy makers to accept the policy implications of scientific research. The naive model holds that networks of scholarly elites can help governments reconcile policy differences in a rational manner. Scientific and professional elites are assumed to be more objective and cosmopolitan and less nationalistic and political than the policy-making elite. The scholarly elites are presumed to have superior knowledge of the essential features of the issues at hand and to be capable of using their knowledge and prestige to persuade policy makers to take a broader view of the issues and to accept solutions which maximize the general welfare. Johnson raises points which run directly counter to each of the key assumptions of the naive model.

The Bellagio group is the more interesting of the two networks Johnson discusses, at least in terms of an attempt to change policy through an elite network. It is worth emphasizing somewhat more than Johnson does that the Bellagio group was a self-conscious attempt by academic economists to influence the reform of the international monetary system. As Johnson notes, there have been three groups in succession within the Bellagio mode. The first Bellagio group consisted of 32 economists who met four times between December 1963 and June 1964 and produced a report entitled *International Monetary Arrangements: The Problem of Choice.* The second Bellagio group was comprised of some of the academic economists from the Group of Thirty-Two and some of the officials directly involved in international monetary reform negotiations. The second group was formally known as the "Joint Conference of Officials and Academics." The Joint Conference met first in December 1964 and may still be holding informal periodic sessions, although I know of no meeting since 1973. (Some of the early working papers of the Joint Conference can be found in Fellner et al., 1966). The third group grew up alongside the Joint Conference and was formally named the "Conference of Practitioners and Academics" and informally called the Burgenstock Group. The Burgenstock Group included academic economists (some from the Joint Conference) and "practitioners" engaged in business and banking, but no government officials, at least initially. The Burgenstock Group was concerned exclusively with the issue of exchange rate flexibility, and it met five times from 1969 through 1971. The papers of the Burgenstock Group were edited by George N. Halm (1970). A follow-up conference at Williamsburg, Virginia in 1973 examined the floating rate experience. Other mutations may have occurred which remain undisclosed.

The first question the naive model of influence suggests is: Does the professional elite constitute a separate, cohesive network? The line separating professional economists from policy makers emerges clearly and sharply both in the organization of the three "Bellagio-type" groups and in Johnson's chapter. As Johnson observes, the Group of Thirty-Two came into being for the purpose of convincing government officials that academic economists could say something consistent and useful about reform of the international monetary system.

However, the experiential gulf between the academic study of economics and the making of economic policy is not so great as the structure of the Joint Conference and the tone of Johnson's chapter would lead one to believe. Within the Joint Conference there were policy makers who were professional economists in their own right, a few having held academic posts prior to entering national or international service. Among the international civil servants who attended the Joint Conference, J.C.R. Dow (a British official of the O.E.C.D.) and Milton Gilbert (an American on the B.I.S. staff) had previously established scholarly reputations. Among the national officials who attended various sessions of the Joint Conference, at least the following had pursued career patterns which combined academic economics with government service: J. Mertens de Wilmars, who served as both counsellor to the National Bank of Belgium and professor of economics at Louvain, and Sir Alec Cairncross, who represented the U.K. at Working Party 3 of O.E.C.D. and has a chair at Oxford. If the professional network is deemed to be broader than the confines of the three "Bellagio-type" groups, the dividing line between economists and policy makers becomes even more obscure. How would Johnson classify, for example, Posthuma of the Netherlands, Cooper of the United States and Schiller of Germany–each of whom held important policy posts and university professorships at various points in their careers?

Then there is the matter of formal educational background. Not too long ago, professors of international relations took hope from the fact that they were training the world's future diplomats and foreign ministers (that was before Henry Kissinger demonstrated that professors of international relations could *become* foreign ministers); much of the indirect influence of academic economists on economic policy surely stems from their role in educating new generations of economic policy makers at universities. I know of no systematic study of where and from whom economic policy makers learned their economics, but that could well be a useful line of research. If Johnson is correct that a new international monetary economics is emerging, perhaps the new academic elite will train a policy-making elite for the 1990s in a more consistent and compatible view of international finance.

Furthermore, there is the matter of staff and consultants and their recruitment. The prestige of the elite network of academic economists gives them considerable influence in the recruitment of staff. Professor Fritz Machlup was asked to recommend a staff economist to study the international monetary reform issue for the Joint Economic Committee of Congress. The staff economist he recommended became a highly effective advocate of flexible exchange rates and probably made Congress more receptive than it would otherwise have been to the float of the U.S. dollar in 1971 and 1973. I suspect that similar examples can be cited of the importance of Oxford and Cambridge economists who acted as consultants to the British government in the 1960s. It could be argued that the economists in the British case did not understand the international monetary problem, but that would not necessarily imply that they lacked influence on policy.

Finally, it should be remembered that there were a great many international conferences on reform of the international monetary system which had nothing to do with the "Bellagio-type" groups, except perhaps to imitate their ambitions. Interaction between academic economists and the policy makers is extensive, even when not direct. Conceptually, one may denote either multiple "networks" or a single "network" encompassing economists of various persuasions as well as policy makers, consultants and staff. Either way, the overall effect is to blur the distinction between a scientific elite and a policy-making elite.

A second general point which contradicts the naive model is that professional scientists can be quite nationalistic. Johnson suggests that as a result of the meetings of the Joint Conference "the academics were forced to abandon many of the national biases in what they had previously believed to be objective analysis of the international monetary problem." Johnson attributes the breakdown of nationalistic orientations more to the unique national background of the participants than to the effect of the conference meetings. He also argues that the second type of network—the "problems" conferences of academic monetary economists initiated by Mundell—was more important than the three "Bellagio-type" groups in broadening the orientation of economists.

Why has the "problems approach" succeeded, if it has, in promoting a more objective, internationalist outlook among international monetary economists? There are a number of lines of speculation which could be pursued. One hypothetical explanation for the apparent nationalism of some Bellagio participants and the internationalism of Mundell's circle may be that the former were more closely involved with policy choices, whereas the work of the latter still appears esoteric from a policy perspective. The test of the influence of the "problems approach" will not be whether the new generations of international monetary economists retain their international orientation, but whether they have any effect on policy.

Third, academic economists have no monopoly on policy-relevant knowledge, although the naive model assumes that they have at least a significant lead. Johnson notes that in the case of the SDR negotiations, the policy makers were considerably ahead of the academic economists:

> The officials' thinking developed faster and more boldly than that of the academics, presumably partly because they were continuously discussing new issues and integrating them into their thinking, in the concrete sense that the officials penetrated the theoretical core of the problem of creating new and genuinely international liquidity, and embodied their understanding in the device of Special Drawing Rights at the IMF, while the academics were still laboring with the conventional concept of an international central bank.

Undoubtedly there are areas in which academic economists can claim prior and superior knowledge. As Johnson suggests, the research done by the networks growing out of Mundell's "problems approach" probably belongs in this category.

Fourth, there is the questionable assumption of the naive model that the scientific elite can persuade the policy-making elite to accept specific policy advice or at least a certain policy perspective. There are basic barriers to the flow of communications between academicians and policy makers even in such comfortable surroundings as an Italian villa. Policy makers are not patient listeners; they are men of action accustomed to receiving their options in brief, crisp memoranda or oral presentations. Academicians are notoriously poor briefers; they have an almost irresistible desire to lecture to officials. The Joint Conference was only partially successful in opening up the flow of effective communication between academic economists and policy makers.

Johnson's chapter suggests not only that academic economists were not persuasive on matters discussed at Bellagio, but that the one subject upon which the economists most wished to lobby the policy makers—flexibility of exchange rates—was excluded from discussion. In fact, I have been told that one economist who did insist upon pressing his views about floating rates at an early session of the Joint Conference was not invited to later sessions—*at the insistence of the national official from his country*. Thus, the policy makers took steps to insure that they would not have to listen to arguments which they found unacceptable.

If the points raised in Johnson's chapter are correct, the naive model of the influence of international professional networks should be discarded.

But what should be put in its place? There are hints in Johnson's chapter of at least three patterns of indirect influence by international professional elite networks upon policy. The first arises through the educational and recruitment processes whereby generations of policy makers acquire their training and rise to responsible positions in government. The question is whether an *international* network creates *international* standards both for education and recruitment.

The second pattern is essentially one of cooptation of academic economists by policy makers. Johnson notes that the first Joint Conference was "successful and interesting enough for the officials to want to repeat it." Actually, my research indicates that the second meeting of the Joint Conference was not held until a full year after the first and was the result of unique circumstances. A subgroup of Working Party 3 of O.E.C.D. (known as the Lawson Group after its Canadian chairman) had been assigned the task of producing a report on the adjustment process to parallel the work of the Group of Ten Deputies on reserve asset creation. The Lawson Group had become deadlocked. Recognizing that without reaching some consensus on the adjustment process it would be politically impossible to get agreement on reserve asset creation, the chairman of Working Party 3 suggested that a "Bellagio-type" study group composed of both academic and officials might prove helpful. Accordingly, the meetings of the Joint Conference held in January and April 1966 were devoted to the issue of the adjustment process. The meetings appear to have provided the national officials with an opportunity to air their views before the academics and subtly compete for scientific legitimization of essentially national viewpoints. After the January meeting of the Joint Conference, the O.E.C.D. staff saw a possibility of reaching agreement and produced a draft report for the Lawson Group. The draft was not entirely satisfactory to the several national governments, but it did stimulate the negotiating process and a report was agreed to and published in August 1966 (O.E.C.D., 1966). The meetings of the Joint Conference were instrumental to the generation of the O.E.C.D. report, although the role of the academic economists may have been largely inadvertent.

The third pattern of indirect influence is represented by the Burgenstock Group, which at most of its meetings included no policy makers at all, although it was the intention of the organizers of the meetings to influence policy toward greater flexibility of exchange rates. The Burgenstock Group took the unusual step of issuing a press release which included a poll of those participating in the meetings. The poll appeared to indicate widespread support among the "practitioners" as well as the academics for greater flexibility of exchange rates. The poll involved a rather artfully constructed set of choices and the results may also have reflected the prior views of the practitioners who attended rather than the persuasive talents of the academicians. In any case, the Burgenstock Group was a public relations success in conferring both academic and business respectability on flexibility of rates. As Johnson observes, the Burgenstock

Group made floating rates easier to accept when they came along, although the Group may not have had much direct effect on exchange rate policy.

These three indirect patterns of influence, (1) education and recruitment, (2) cooptation, and (3) public relations, are only illustrative. They do imply, however, that when the international professional elite network of economists goes beyond its traditional and glacial shaping of economic theory, the interaction between academics and officials becomes highly political and direct influence by academic economists on public officials is rare. That should not be surprising. Policy makers are unlikely to be swayed by either the cosmopolitanism or the superior knowledge which academicians may, on occasion, display. The officials meet each other far more often than do professional economists, and in equally pleasant surroundings. Officials have no great difficulty grasping the technical features of issues and are usually able to agree on what the choices are. The basic disagreements between officials reflect matters of political choice, which is why international networks of professional economists like the Bellagio and Mundell groups are only marginally influential.

REFERENCES

FELLNER, W., F. MACHLUP, R. TRIFFIN et al. (1966) Maintaining and Restoring Balance in International Payments. Princeton, NJ: Princeton Univ. Press.

HALM, G. N. (1970) Approaches to Greater Flexibility of Exchange Rates: The Burgenstock Papers. Princeton, NJ: Princeton Univ. Press.

Office of Economic Cooperation and Development (1966) The Balance of Payments Adjustment Process. Paris: Author.

COMMENTARY

Leonid Hurwicz

The two chapters of Part II by Harry Johnson and Lawrence Klein are extremely valuable for several reasons. They provide factual information in the respective domains of their studies that can only come from persons actively engaged in, and indeed generating, the international activities they describe. They provide, or at least suggest, an analytical framework that can be used for an evaluation of effects of the scientific interactions they deal with. Also, while "accentuating the positive," they point out gaps, imperfections, and negative effects.

While both deal with activities of the economists on the international stage, their topics are so different in nature that to a considerable extent each must be examined within its own framework. Yet in a broader perspective they help us understand different facets of the "economic scientific component" of the international interaction system.

One could think of this system as an expanded form of Klein's LINK model. In this expanded model there is a submodel for each country or group of countries and certain linkage relations between them. The variables entering the equations of the model are typically "economic" in nature—prices, exports, imports, and so on. However, hidden among the dots of (say) his equations 6 and 7, there can be, in principle at least, "noneconomic" variables, for instance those determining national policies. It is not necessary that the noneconomic variables be exogenous. Political and social indices might well be introduced as causal factors in some of the equations, but they might appear as dependent variables in other parts of the system, thus becoming endogenous.

Among these noneconomic variables are the cultural, and, in particular, scientific and informational, activities that influence behavior in the economic sphere. From this point of view, the networks described by Johnson may be regarded as such variables. (I should emphasize that the term

AUTHOR'S NOTE: The very interesting study by Mesarovic and Hughes had not reached me at the time this was being written. I did not consider it appropriate to add comments which had not been presented at the conference.

"variable" is used abstractly and need not be confined to measurable or quantitative phenomena.) Presumably, the ultimate objective of research initiated in Johnson's chapter is to "estimate" the causal relationship between the activities of such networks and the resulting policy decisions. But this is, of course, an ambitious objective. What Johnson concentrates on instead are the interrelationships between the intellectual activity variables in different countries and between different strata (governmental versus bureaucratic) involved in the process.

To the extent that the causal direction goes both ways, there is endogeneity within the intellectual sphere. But one need not be a devotee of historical materialism to ask whether there are important causal relations from the sphere of "physical" phenomena (whether they be political, economic, or social) to those in the realm of scientific activity. The emergence of the LINK project, due in part at least to the specific needs of the present era, and made possible for the first time in history by the development of modern high-capacity high-speed computer technology, is an important example. The sequence of events detailed by Johnson leading from a change in the U.S. stance on monetary reform to the creation of the Bellagio group is another.

Let me return to the phenomenon of interrelations between intellectual activities in different countries. To assess the role of groups such as those described in Johnson's chapter, a somewhat wider perspective seems appropriate. The flow of economic ideas from country to country is an old phenomenon; one need only recall Adam Smith's internationally influential classic. Personal contacts, group formations, and conferences do facilitate and intensify transmission of information, ideas, and analytical frameworks. We should not ignore the fact, however, that historically the printed word has had powerful impact, especially in the area where economics and ideology meet (Smith, Keynes, Marx, Proudhon, Hayek, Friedman) as well as in propagating analytical frameworks. Furthermore, beyond the printed word (books, scientific journals, and so on), there are such forms of international transmission of ideas as graduate study abroad, and finally the regular meetings of international scientific societies. Here we do have personal contact and both formal and informal group dynamics, but not nearly as goal-oriented as, for instance, the Bellagio group.

Still, judging by personal experience of meetings of one such organization (the Econometric Society), one should not underestimate the extent to which techniques, ideas, and analytical models are disseminated through such meetings. Indeed, one can observe there phenomena of transition from the accumulated wisdom of one generation to the unveiling of new points of view, technical competence, and so forth, of another—although there are classical examples (Frisch, Tinbergen) of scholars who remained in the "young" generation long past the chronological entitlement; examples of the reverse phenomenon are not unknown.

Johnson describes in vivid color how monetarist ideas are propagated through the Money Study Group in the United Kingdom and the Konstanz

Seminar. But some decades earlier, Keynes' model received wide acceptance among economists throughout the world, without (as far as I am aware) comparable institutional arrangements, and despite (or because of?) a torrent of criticism from highly prestigious economists of that era. Far be it from me to deny that the formation of groups such as those described by Johnson makes a difference, but it is not yet clear to me just what this difference is. Is it the speed of propagation? Is it the influence on those close to the policy decision-making process? Further research designed to help answer such questions would seem a proper follow-up to Johnson's study.

In tracing the process of transmission of the economists' ideas, it is important, in my view, to distinguish the different "intellectual products" that are being transmitted. These "products" include: empirical information (both "raw"—e.g., price data—and "processed"—e.g., estimates of structural coefficients in various interaction equations); analytical frameworks and models; subjective judgments concerning facts; values (e.g., attitudes toward planning, decentralization, consumer sovereignty, the relative importance of efficiency versus distributive shares).

Insofar as we try to analyze the very recent cases of transmission of ideas, it is naturally difficult to separate oneself from one's own convictions and value judgments. In his section on "The Problems Approach," Johnson's language is not used to conceal his feelings. He speaks, for example, of Mundell's early ideas being "perverted to the service of an anti-American European nationalism of a somewhat repulsive kind." Among positive accomplishments of the "networks" he lists "the recognition of the reality of world inflation in a system of fixed exchange rates, and its basic origins in excessive monetary expansion, particularly in the United States as reserve currency country of the system." Elsewhere in this section, he speaks of "the stultifyingly deeply entrenched Oxford-Cambridge Keynesian approach to macroeconomic policy." The last two comments are shared by many respected economists, but would be disputed by many others. In analyzing the role played by the networks, therefore, it would be desirable to separate the question of how ideas spread and influence policy from the question of their correctness or validity, at least where there is significant controversy. (Or am I merely proving that the statement attributed to Robert Roosa is correct? See Johnson's section "The Bellagio Group.")

Similar issues arise in the more purely methodological sphere. Personally, I am in sympathy with the utilization of the hypothesis of maximizing behavior. However, I do not share the view that its rejection necessarily leads to the acceptance of government intervention in the determination of exchange rates. There are alternative behavioral hypotheses that merit consideration. Should one of them be accepted, it might turn out to be compatible with noninterventionist policy conclusions.

Questions of methodology and its policy implications also arise in connection with work whose basic objectives are empirical, such as project

LINK. As Klein notes, LINK does not try to standardize models, but merely to connect them in an appropriate way. Now it is in general the case that the so-called identifying assumptions made in one part of a simultaneous equation model affect the estimates obtained for the parameters in others (whether this is so here, I would be interested to find out). Hence, it can make a difference whether the assumptions about economic behavior are acceptable. Even if there were no indirect impact on parameter estimates, incorrect values of a given country's parameters would warp the projections of consequences of alternative economic policies pursued by other countries or by international bodies. Hence there are some dangers in the laissez-faire attitude of LINK toward individual country models. One should recognize, however, that—in the short run, at least—there may be no feasible alternative.

To what extent do large scale econometric models reflect behavioral characteristics (such as various forms of maximizing behavior)? If they did not, their validity as policy formation aids could be questioned. Are they "stultifyingly Keynesian"? Typically, they do embody important elements of structure derived from the early Keynesian models. On the other hand, provided the a priori restrictions are not unduly narrow, they can lead to parameter values corresponding to "anti-Keynesian" hypotheses concerning the relationship between income and consumption—or with regard to the relative effectiveness of monetary versus fiscal policy. Nevertheless, alternative model formulations—not equally acceptable as statements of fact to various economists—can affect explanations of past events, forecasts, and policy recommendations. It is important, therefore, to educate the prospective user—domestic or international—about the subjective component of such systems and about the sensitivity (or, it is hoped, robustness) of conclusions. In any case, here again we are witnessing not merely the transmission of techniques and information, but also of beliefs and judgments about which reasonable people might differ.

Having stressed the subjective element, let me say that modern economics is characterized to a much greater extent than its ancestors by a common technical framework and internal ability to communicate, as well as the ability to separate a priori beliefs and value judgments from the common "scientific" component. This, of course, facilitates international interaction at the level of ideas. However, there has been much less development of such a common core in analyzing the workings of different institutional economic systems, e.g., capitalism versus socialism. Fortunately, the gulf is being bridged in certain selected areas. The conference of June 1976 in Moscow at which both empirical and analytical papers were presented by U.S. and Soviet economists (including Klein, Modigliani, and myself) showed a significant improvement in mutual technical comprehension, increased interest in the "other side's" approach to problems, a good deal of agreement on certain empirical and analytical techniques, but, of course, also a great divergence of viewpoints. It is a tantalizing question whether such exchanges will have significant effect on the development of economic science in either country or on policy.

PART III:
POLITICAL IMPACT OF ISPAs

7

COSPAR AND SCOR: THE POLITICAL INFLUENCE OF THE COMMITTEE ON SPACE RESEARCH AND THE SCIENTIFIC COMMITTEE ON OCEANIC RESEARCH

Edward Miles

While there exist several general analyses of International Nongovernmental Organizations (INGOs) and their roles in the international system (White, 1951; Lador-Lederer, 1963; Smoker, 1965; Angell, 1969; Skjelsbaek, 1971), we have had very few comprehensive evaluations of specific INGOs assessing the difference they might make in the conduct of world politics.[1]

At the level of the international system, scientifically-oriented INGOs—one type of ISPA—and primarily those encompassed within the International Council of Scientific Unions (ICSU), perform the following functions:[2]

(a) They are for organizing and maintaining international contact among scientists, and this is closely related to the presentation and stimulation of research and diffusion of findings.

(b) They are for the planning and organization of cooperative international research projects like the International Geophysical Year

AUTHOR'S NOTE: The author is immensely indebted to Professor Roger Revelle for information on the prehistory of SCOR.

(IGY), the Global Atmospheric Research Program (GARP), the International Year of the Quiet Sun (IQSY), and the International Biological Program (IBP).

(c) They perform as centers for the aggregation and articulation of interests in particular disciplines and lobby at both national and international levels.

(d) They are used as repositories of highly technical advice on particular issues at the international level.

(e) They seek to ensure comparability of data within particular fields.

(f) They are beginning to engage in scientific and technological forecasting.

Within ICSU, most of the task expansion which has occurred has been the result of new fields or problems of research gaining recognition at the international level. Interested scientists then organize transnationally to perform the functions described above and this leads to the creation of Scientific or Special Committees like the Committee on Space Research (COSPAR) or the Scientific Committee on Oceanic Research (SCOR), the Scientific Committee on Antarctic Research (SCAR), and the Special Committee on the International Biological Program (SCIBP). For ICSU, therefore, the major organizational effect of new research problems is expansion of the scope of existing tasks expressed in terms of increasing horizontal differentiation. Some ICSU members have long played major advisory roles to intergovernmental organizations (IGOs). It may be that they will begin to do this for governments of developing countries as well, but there are only a few tentative hints of this possibility at present.

The purpose of this chapter is to provide a comparative analysis of the political links and behavior of two relatively recent members of the ICSU family. COSPAR and SCOR were chosen because: (1) they are both direct progeny of the International Geophysical Year (IGY) experience; (2) both deal with the world's commons, i.e., outerspace and the oceans, but the political constellations surrounding them are different, given differences in the distribution of capabilities to explore and exploit the medium and, therefore, differences in the diversity of benefits (or payoffs) to be gained; (3) both have scientific as well as policy advising functions; (4) both penetrate and are penetrated by governmental and intergovernmental structures and, in that sense, understand the nature of transnational politics and deliberately seek to play that game (Keohane and Nye, 1971).

The intent of the comparison is to assess the *political significance* of the range of organizational behaviors exhibited by COSPAR and SCOR. Political significance is defined in terms of actual and/or potential impact on governmental decisions with respect to courses of action, i.e., policy.

We will focus on five specific kinds of behavior. These are actions: (1) to create new national and/or international programs, organizations or networks; (2) to rearrange existing national and/or international programs, organizations or networks; (3) to increase budgetary allocations to particular scientific disciplines; (4) to advise on controversial choices by evaluating the probable consequences of alternative causes of action; and (5) to protect the interests of member scientists vis-à-vis proposed governmental decisions.

The behavior of COSPAR and SCOR suggests that two avenues are the primary ones for facilitating the exercise of influence by international nongovernmental scientific organizations. This first is provided by their role as a forum. This is important for two reasons: (a) as a sounding board for new ideas and for lending credibility to different and sometimes unconventional proposals; and (b) affording skilled leadership the opportunity to reduce a wide variety of demands and preferences into a coherent program which senior, authoritative scientists then urge upon their own governments. At the same time, the same demands, if appropriate, can be made of different intergovernmental organizations (IGOs) in a coordinated strategy. The second avenue of influence lies in technical expertise which allows some small groups of specialists to define the *outer limits* of the range of political choices which are congruent with technical criteria.

COSPAR

The primary purpose of COSPAR as stated in its 1960 charter is "to further on an international scale the progress of all kinds of scientific investigations which are carried out with the use of rockets or rocket-propelled vehicles." However, the organization limits itself to basic research and does not deal with "such technological problems as propulsion, construction of rockets, guidance and control."[3]

As a Scientific Committee and not a member union of ICSU, COSPAR has two types of members: (a) one representative of each national scientific institution actively engaged in space research and already a member of ICSU; and (b) one representative of each national scientific union belonging to ICSU and desiring to participate in COSPAR's activities. There are 34 national members of COSPAR and twelve international scientific union members.

We should note that while COSPAR is formally a nongovernmental organization, governments, particularly those of the Soviet Union and the United States, can and do exercise considerable influence on certain organizational programs. We do not suggest that this dimension of governmental penetration is necessarily a disadvantage in the conduct of programs. In fact, it may have been a source of considerable informal influence. In the words of one NASA official, "COSPAR brings space scientists together in an environment that minimizes political considerations, yet permits responsible and authoritative planning" (Frutkin, 1965; see also Odishaw, 1962). This was the major lesson of the IGY which has been repeated several times; the nongovernmental setting gives governments the greatest flexibility when large-scale international scientific programs must be planned requiring significant financial commitments and allows maximum insulation against infection by external political conflicts.

Members of COSPAR tend to see its primary task as being the exchange of information which takes place within the Working Groups. The effects of such exchange and formation of collective opinion may be of special significance to *national* programs of space research, especially in the setting of priorities and allocation of resources (Frutkin, 1965; COSPAR, 1971; Roy, 1969). COSPAR has also acted since 1958 as the major international appraisor of potentially harmful experiments in space, and it provides technical advice on request of the UN General Assembly and various Specialized Agencies.

COSPAR has performed evaluations of controversial governmental decisions on two occasions. In early 1958, the U.S. National Academy of Sciences had become sufficiently concerned about the probability of radioactive and biological contamination of the moon, Mars, and possibly Venus that it passed a resolution urging scientists to plan "lunar and planetary studies with great care and deep concern so that initial operations do not compromise and make impossible forever after critical scientific experiments" (COSPAR, 1965: 88). The Academy also urged ICSU to consider this question and an Ad Hoc Committee on Contamination by Extra-Terrestrial Exploration (CETEX) was appointed by The Bureau of ICSU in March 1958. This committee prepared a report submitted in 1959 containing a number of general criteria, since the group did not think it possessed sufficient scientific and technological information to formulate a specific code of conduct (COSPAR, 1965; Johnson, 1964). It recommended, however, that this be done as soon as possible.

On submitting this report in May 1959, CETEX was dissolved; but its task devolved upon the newly-created COSPAR when, in September 1961, the ICSU General Assembly invited COSPAR to examine the question of harmful effects of space experiments and the probability of harmful effects inherent in different kinds of experiments. To keep this problem under continuing review, COSPAR, in May 1962, created the Consultative Group on Potentially Harmful Effects of Space Experiment; and in May

1964 a study group met in Florence to consider standards for space probe sterilization. This study group consisted of scientists from Sweden, the United States, France, Belgium, the USSR, and the United Kingdom, and its recommendations were later adopted by the COSPAR Executive Council (see Sneath 1968). The proposed standards are that:

> the probability of a single viable organism aboard any spacecraft intended for Mars landing be less than 1×10^{-4} and that the probability of accidental impact over a period of decades by each unsterilized flyby or orbiter be 3×10^{-5} or less [Sagan et al., 1968].

These standards at first were not accepted and implemented by either the United States or the Soviet Union primarily for two reasons. The cost of doing so was estimated to be quite high; and there was no great international pressure on the superpowers to implement the standards, because they were the only ones involved. But, over time, the COSPAR recommendations have been influential because there is a subtle and long-term relationship between the work of the COSPAR Consultative Group and the recommendations of participating scientists to their own governments.

For example, the author interviewed the appropriate officials of the U.S. National Aeronautics and Space Administration (NASA) on this question in the fall of 1968; and they were hostile to COSPAR's recommendation and openly critical of Sagan, Levinthal, and Lederberg for their involvement. However, a more recent publication of NASA relating to this question takes a completely different point of view. The recommendations of the COSPAR Group are characterized as a "milestone" because it provided an analytical rationale and, for the first time, "international agreement on quantitative objectives in terms of probabilities of events which characterize planetary contamination" (NASA, 1975: 23).

The Consultative Group had also been involved in another contentious issue concerning Project Westford carried out in 1963 (Skolnikoff, 1967; see also COSPAR, 1961a, 1961b). This project called for the orbiting of a large number of copper dipoles to facilitate "over the horizon" communication, but it raised fears among radio astronomers in the United States and elsewhere that harmful and possibly irreversible effects would occur. While the U.S. government did not formally seek COSPAR's approval for this, the issue was considered by the Consultative Group because its scientific implications were worldwide. The result of the investigation was that these dipoles would have no lasting adverse effects. It is possible that whenever such issues arise affecting the interests of large numbers of scientists in several countries, COSPAR, if it survives, will continue to play the role of authoritative appraiser.[4]

COSPAR's joint involvement in the politics of frequency allocations for radio astronomy has been extremely effective.[5] The intergovernmental Administrative Radio Conference of the ITU held in Geneva in 1959 was the first occasion on which bands for space services were allocated. At this time, the radio astronomers managed to receive some bands but because they were required to share these frequencies with other users they considered their allocation to be most inadequate, given the vast energy sources and transmission distances with which they had to contend (ITU, 1963). In fact, it was estimated at the time that the United States took the minimal and rigid position for radio astronomy that it did because the value of that portion of the spectrum desired by the scientists on an exclusive basis would have amounted to about $100 million a year to U.S. communications companies (Wenk, 1960). In addition, the U.S. Department of Defense was concerned about the security aspects of allowing radio astronomy to compete for frequencies in which it was interested.

Even as the Conference was in session, U.S. radio astronomers met with representatives of the Departments of State and Defense and the Federal Communications Commission in an unsuccessful attempt to change the U.S. stance. This had implications, of course, for radio astronomers everywhere, because if the United States could have been persuaded to change its position this would have facilitated attempts by the radio astronomers to obtain adequate frequency allocations.

As a result of this experience, scientists from three ICSU members, the International Union of Radio Sciences (URSI), the International Astronomical Union (IAU), and COSPAR met in 1960 and established the Inter-Union Committee on Frequency Allocation (IUCAF), which was recognized by the ICSU Board in October of that year (COSPAR, 1961a). The purpose of IUCAF was to aggregate and represent the interests of scientists involved in various aspects of radio research, radio astronomy, and space research. This body then made a series of recommendations to the IFRB for the 1963 Conference (ITU, 1963).

In addition to this, COSPAR had, in January of 1960, decided to establish links with the International Telecommunication Union (ITU). This was done when the ITU agreed to extend consultative status to COSPAR for the purpose of participating in the work of the International Radio Consultative Committee (CCIR), in particular with Study Group IV on Space Systems and Study Group VI on Ionospheric Propagation (COSPAR, 1960b). Several years later, not willing to leave anything to chance for the second space conference planned for 1971, IUCAF urged those national members of IAU, URSI, and COSPAR which had not already done so to establish direct contacts with their respective national radio administrations to ensure that "they have channels whereby effective consultation will be possible if and when needed" (COSPAR, 1964: 20).

As a result of all this lobbying, IUCAF obtained frequency bands for radio astronomers allocated on an exclusive or shared basis and 22 specific bands allocated to space science research, exclusive of communication and satellite tracking. This organization also continues to monitor the actual usage and effects of frequencies allocated to radio astronomy and submits its observations to the IFRB through national delegations to the ITU (COSPAR, 1966).

THE POLITICAL SIGNIFICANCE OF COSPAR

COSPAR has had very little direct involvement with the negotiations taking place with the UN's Committee on the Peaceful Uses of Outer Space since 1961. While participants aver that it has had some impacts on the development of national space programs, even of the superpowers, this is difficult to document directly. As indicated previously, however, this was true on the issue of contamination of planets. It does not appear to have had much direct impact on domestic allocations to space research, although this, too, is difficult to assess.

What, then, has been the primary political significance of COSPAR? Two links appear to be crucial. At the international level, COSPAR serves to articulate and aggregate the interests of a wide variety of space scientists and can effectively mobilize these scientists on any issue which is seen to be important by a large number from several countries. But perhaps an even more important link exists in the fact that many of COSPAR's members, and particularly the representatives serving on the Executive Council, are senior scientific advisors on space questions for their own governments; and the informal discussions which take place in COSPAR are sometimes important for governmental decisions. We do know that it is not at all uncommon for scientists from both developed and developing countries to use membership in international organizations as a means of increasing their own budgetary allocations nationally to engage in programs in which they are interested. To summarize the responses of two members of the Executive Council, one of the primary political functions of COSPAR is to educate governments on the question of space science, to exert pressure for converging action, and to provide information "based on the common feeling of the scientists."

SCOR

Created in 1957-1958 as a new kind of mechanism within the ICSU family, SCOR has had a major impact on intergovernmental and international nongovernmental decisions to create new programs, organizations, and networks dealing with marine science. In addition, SCOR has had significant influence both on national decisions to increase budgetary allocations to marine science in developed as well as developing countries and on the development of programs within the Intergovernmental Oceanographic Commission (IOC) of UNESCO. On the other hand, in a strikingly different approach from the radio astronomers, SCOR has been unable to organize an effective effort to protect the interests of marine scientists within the Third United Nations Conference on the Law of the Sea (UNCLOS III).

The terms of reference for SCOR have undergone many changes since 1957; but several of them remain the ones usually cited for member unions of ICSU. These include: promotion of scientific advance via clear formulation of scientific problems to be investigated, standardization of methods of investigation and measurement of phenomena, and exchange of information via meetings and publication of reports. In addition to these, one of the original purposes of SCOR was to provide a forum for *interdisciplinary* discussions of oceanography, something hitherto not possible on a systematic basis within ICSU, given the dominance of the disciplinary Unions.[6] SCOR also later assumed an important role as advisor to the Intergovernmental Oceanographic Commission of UNESCO on a variety of marine science-related issues (Wooster, 1969).

The early effectiveness of SCOR as a transnational actor is best demonstrated by looking at three events: (a) the creation of SCOR, (b) the launching of the International Indian Ocean Expedition, and (c) the creation of the Intergovernmental Oceanographic Commission in UNESCO.

The prehistory of SCOR goes back to 1946, though the most crucial thrust dates from 1954. The membership of the Joint Commission for Oceanography at that time included several of the major statesmen of oceanography, in particular Anton Bruun (Denmark), George Deacon (U.K.), and Roger Revelle (U.S.). The major accomplishment of this Commission was the establishment of a journal for deep sea research. The Commission was chaired by John Weisman of the British Museum (Natural History). As a result of internal conflicts between Weisman and Deacon (National Institute of Oceanography), it had a very short life, also being abolished in 1954. Bruun, Harald Sverdrup (Norway), Deacon, and Revelle were asked to consider what should replace it, but nothing specific came of this.

While these events were taking place in ICSU, another sequence was initiated in UNESCO almost simultaneously; and, not surprisingly, the links between ICSU and UNESCO were made almost immediately. At that time, Pierre Auger was Assistant Director General for Science in UNESCO, and he thought that the organization should be involved in oceanography. He was able to secure a small annual budget of about US $30,000 and formed the International Advisory Committee on Marine Sciences (IACOMS) in 1955. IACOMS consisted of nine members and these included some ICSU people, particularly Revelle.

The cost of an annual meeting for IACOMS amounted to approximately $10,000, so the question was raised concerning the use to which the remaining funds should be put. It was at this point that planning for the IGY provided the major stimulus. Oceanography had only a small part to play in the IGY, since synoptic observations were not seen to be as necessary as with solar research or determining changes in the earth's magnetosphere. While there was strong interest in the United States to set up tide gauges to study annual global variations in sea level, the most important joint effort in oceanography and meteorology involved the setting up of stations to study carbon dioxide distribution in the atmosphere and sea surface waters. The planning for these experiments was especially important in a serendipitous way since for the first time; and, in spite of the intense political hostility existing between the U.S. and the USSR, it brought Soviet and U.S. oceanographers together during 1954-1955.

We return now to the ICSU track in 1956. Lloyd Berkner (U.S.) was a major figure in ICSU at the time. He had been president and was then Vice President of the Organizing Committee for the IGY (CSAGI). He approached Revelle saying that ICSU wanted to continue its activities in oceanography and suggesting the appointment of a special committee to define a longer-range program. Revelle went to Europe to consult with Deacon, Bruun, and Boehnecke (Federal Republic of Germany and later to be Secretary of SCOR). These included other influential scientists like Columbus Iselin (U.S.), M. N. Hill (U.K.), Luis Capurro (Argentina), N. K. Pannikar (India), and K. Sugawara (Japan). Pannikar and Sugawara were also members of IACOMS. A meeting was arranged in Woods Hole in August of 1957, and SCOR was born with Revelle as its first President. It was accepted as a Special Committee by ICSU in 1958 with a stipulated life of five years to define a long-term program. After that time, it became a permanent Scientific Committee of ICSU in 1963.

The forging of the formal link between SCOR and UNESCO clearly required the dissolution of IACOMS. This was resented at first by Pierre Auger, since IACOMS was appointed by him and was therefore a significant potential source of influence. However, since the SCOR coalition consisted of very influential scientist/administrators from advanced oceanographic countries, the obvious strategy was to defeat the UNESCO bureaucracy by changing the official positions of the most influential countries. In this way, IACOMS was eventually suppressed (1960), and

SCOR became the official advisor on marine science to UNESCO on a contract basis.[7]

There is a direct link between the creation of SCOR, the establishment of the IIOE as its major focus, and the creation of the IOC as the primary intergovernmental mechanism and pressure group for marine science. The Executive Board of ICSU, at its meeting in June 1956, in authorizing the SCOR meeting for 1957, had given it the following terms of reference:

> The task of SCOR would be to frame a five-year programme for a cooperative effort in oceanic research, to initiate action on a national basis through the National Members of ICSU, and to organize such meetings and to authorize such publications as may be necessary for the completion of the programme.

At the first meeting in Woods Hole in 1957, SCOR defined those long-range problems which were seen to be critical: (a) the use of the deep sea as a receptacle for industrial wastes, (b) the ocean as a source of protein, and (c) the role of the ocean in climatic change. "For all three of these problems, greater knowledge is required of the rate and character of the exchange between the deep and surface waters of the oceans" (Special Committee on Oceanic Research, 1957: 2). This formulation was very ingenious, since it sought to package a problem of great and immediate scientific interest—i.e., coordinated observation and measurement of the deep ocean—in terms of its anticipated social utility. The scientific objective was the mounting of a major effort to understand the relationships of the flora and fauna of the deep ocean with the dynamics of the upper water layers. In addition, the opportunity would be provided to study closely the region below the thermocline down to the deep sea floor, including a study of the floor itself. The drawback was that this was an undertaking of enormous expense, since it would involve a large number of heavily equipped ships for long periods of time. But economies of scale could be derived if several countries contributed significant amounts of ship time.

It was therefore in this context that Columbus Iselin, then Director of Woods Hole Oceanographic Institution, proposed, and SCOR accepted, that a large multination, multiship cooperative investigation of the Indian Ocean be attempted. The Indian Ocean was chosen because it was the area about which least was known at the time. The program again was sold partly in terms of the benefits it would provide to poor countries on the rim of the Indian Ocean. It is questionable how much of this was actually realized, but the program proved to be of considerable benefit to oceanography in developed countries and in at least one developing country: India. New vessels were built and budgetary allocations to oceanography significantly increased at national levels. The effects of the social utility arguments were to return to haunt oceanographers from advanced countries fifteen years later in the context of the Third United Nations Conference on the Law of the Sea. There, developing countries worked to extend control over marine science conducted within 200 miles of their coasts, precisely because of its presumed social utility. Now, however, oceanographers from developed countries were insisting that there was only a tenuous connection between basic science and its societal payoffs.

Having launched the IIOE, SCOR found that its scope soon surpassed the resources of the scientists themselves. Government commitment and support were absolutely necessary if the program were to be implemented.[8] It was out of this necessity that the IOC was born and the idea came once again from Revelle. In fact, Revelle was pursuing these objectives both in the United States and internationally prior to 1960, but they were not accepted by either the Europeans or the Japanese. But pressure for this alternative was persistent from within the U.S. scientific community, which took the initiative and played effective transnational strategies.

First, Revelle and his colleagues were convinced that there was a need to develop international cooperation in marine science at two levels: among scientists (i.e., those who actually did the work) and among governments, since the support of the latter was essential. The time in the United States was favorable, since it was clear that the Soviets had made major commitments to developing a large submarine force by 1958. The event of *Sputnik 1* also enhanced the urgency of an effective U.S. response. There was consequently a coalition forged between the marine scientists led by Revelle and Wilbert H. Chapman in the National Academy of Sciences, National Research Council, those within the federal bureaucracy pursuing the same objectives and organized within the Interagency Committee of Oceanography chaired by James Wakelin, then Assistant-Secretary of the Navy for Research and Development, and supporters of this enterprise within the U.S. Congress. This coalition enabled the U.S. National Committee for SCOR (within NAS/NRC) to speak with a single voice and to commit U.S. resources. The reports of the NAS Committee on Oceanography were actually very important in stimulating Congressional action and support, not only in terms of budgetary allocations but even in the establishment of Subcommittees on Oceanography in both the U.S. House and the Senate.

On the nongovernmental side, Revelle, Iselin, and Chapman were the crucial mobilizers. Revelle and Iselin presided over the two largest research institutions in the United States (Scripps and Woods Hole), while Chapman personified the interests of fisheries scientists. In this, he was effectively supported by Milner B. Schaefer (U.S.) and Cyril Lucas (U.K.). This collaboration muted a certain tension both in the United States and internationally. The reason for this tension was that fisheries scientists felt at a disadvantage to physical oceanographers in the domestic/international competition for funds. SCOR at the time was dominated by physical oceanographers and, with its links to UNESCO, loomed even larger on the horizon. The fisheries scientists, partly as a result of this and partly for internal reasons within FAO, later organized the nongovernmental Advisory Committee for Marine Resources Research (ACMRR), attached to the Assistant Director-General for Fisheries in FAO. But relations were also bad between the Directors-General of UNESCO and FAO as a result of competition for control of marine science programs. Chapman played a crucial conciliatory role nationally and transnationally, because he was convinced that major advances for fisheries science would come from physical oceanography, and for this reason played both SCOR/UNESCO and ACMRR/FAO games, occasionally helping to merge them on particular issues.

Given this situation, it was not difficult for the scientists to get official U.S. government support for the creation of IOC. The position was worked out during 1958-1960 in the Interagency Committee on Oceanography, and Wakelin chaired the U.S. delegation which negotiated the birth of IOC in Copenhagen in 1960. While the Copenhagen meeting was formally the most important, again it was not substantially so, because in 1959 a small group consisting of Revelle, Deacon, Vladimir Kort (USSR), Bruun, and others met in UNESCO to draft the Charter for IOC. The concept agreed upon was that IOC would be under the umbrella of UNESCO but would be an independent commission. The Soviet Academy of Sciences was extremely interested in IOC because at that time, and for several years afterwards, their only possibility of interacting with oceanographers from other countries was within the IOC. Additionally, the budgetary significance of international commitments was as high for Soviet oceanography as for U.S. and Western Europe oceanography.

Two conceptual mistakes were made at this time, the effects of which returned to haunt the IOC. The first was that membership was defined in terms of countries doing oceanography. The IOC was therefore conceived as: (a) an exclusive club of scientists; (b) who wanted to do research in the deep ocean via large-scale, multiship, multination cooperative expeditions. The first conception was to earn IOC the enmity of the developing countries and considerable hostility at the Law of the Sea Conference, resulting in continuing refusals to allow significant expansions in IOC's tasks. The second was to focus IOC's scientific program too closely on cooperative expeditions. Once the perceived utility of these diminished significantly, there was little or nothing IOC could offer developed countries as an incentive for continued commitment.

After the creation of SCOR, the launching of the IIOE and the creation of IOC, two tasks remained. The first was formally to change the Constitution of SCOR to allow it to play a continuing advisory role to UNESCO/IOC and thereby to affect the elaboration and implementation of governmentally supported programs in oceanography. The second was to pass over administrative control for the IIOE to the IOC. These actions were formally accomplished by early 1962 (SCOR, 1962a; 1962b). The international system for marine science thereby created consisted of a tightly merged governmental/nongovernmental arrangement with the transnational player retaining major influence in the programs and activities of the governmental actor and deriving significant income ($10,000 per year) as a result. The complex coordination that would be required was first to be provided by Warren Wooster of Scripps Institution, later to become one of the prime movers of SCOR.

THE IMPACT OF SCOR ON SCIENTIFIC PLANNING WITHIN THE IOC

The impact of SCOR on IOC-sponsored cooperative expeditions and on comprehensive scientific planning has been considerable. This is not surprising, given the extensive interpenetration of governmental/nongovernmental networks for international marine science. In addition, size is an important factor. Both networks have been small, facilitating the emergence of a club in which everyone knew everyone else personally and, over a long period of time, saw each other frequently. The nature of the advisory relationship between SCOR/IOC and SCOR/UNESCO also served to reinforce this close relationship.

Assessments of the degree to which SCOR has actually affected IOC performance must discriminate between two levels of impacts. The first is measured by SCOR impacts on IOC program decisions, and here the effect has been great. The second, however, is measured by IOC performance once a program decision has been made. This leads to a distinction between IOC performance on programs of cooperative investigations, which has been variable, and IOC performance on attempts to produce comprehensive scientific plans like the General Scientific Framework (GSF), the Long-Term and Expanded Programme of Oceanic Research (LEPOR), and Global Investigations of Pollution in the Marine Environment (GIPME). The latter have been relatively futile exercises in a direct sense, though they have occasionally had fruitful results indirectly for oceanography in different countries.

Wooster (1973) has already recounted the history behind the development of the GSF, LEPOR, and GIPME, so we shall not repeat that here. There are, however, several interesting reasons which explain the manifest futility of large-scale scientific planning in the IOC. The first, and most devastating, is that science simply is not done that way. It is almost always impossible to get consensus on a fixed set of priorities in a decisional framework which is based on voluntary participation. The problem, therefore, always resolves itself into one of assigning priorities in a shopping list which itself cuts across several different disciplines.

This is actually seen very clearly by the participants themselves. For instance, as seen by Wooster: "The fundamental paradox is that governments require priorities for the allocation of funds but scientists don't because they have no criteria to apply and short-run payoff is not considered sufficient. . . . It is the random choices of scientists which explain scientific advance."[9] Another U.S. scientist, present at a meeting in Halifax in 1963 where the GSF was discussed, put the problem this way: "My prevailing feeling for the two-day meeting was that most of the people present were reluctant to draw up any long-range plans in which they might later find themselves irretrievably involved" (SCOR, 1963). Finally, the British National Committee for Oceanic Research (1963: 1), replying to SCOR Circular 12/1962 on the GSF, began by saying: "The British National Committee sees the advantage of planning surveys but considers that attempts to lay down a rigid framework for comprehensive study might do little to further the development of the subject and might hinder the recognition of new, unorthodox approaches, and the emergence of new ideas."

Given these attitudes, which were clear as early as 1962-1963 and have remained unchanged, why has the IOC persisted in pursuing the Holy Grail? The reason, it seems, is that such programmatic statements to which governments formally commit themselves often make it easier for scientists in different countries to get money from their own governments to work on problems in which they are interested.

Another important range of political behavior of the SCOR Executive Committee is that, on specific issues of interest to them, the members of this body, who are simultaneously chairpersons of their countries' delegations at the IOC, function as a coherent transnational delegation. On several occasions, for instance, a particular member of the SCOR Executive Committee has tried unsuccessfully to get his government's delegation to introduce a particular resolution. This does not present any problem at all, since he merely goes to one of his colleagues leading another delegation and has him do it.

THE RESPONSE OF SCOR TO THE THIRD CONFERENCE ON THE LAW OF THE SEA

Compared to the performance of the radio astronomers when their interests were threatened on the assignment of frequencies, this is a case of dogs which did not bark in the night.

As early as September 1970, before the UN Committee on the Peaceful Uses of the Sea-bed and Ocean Floor Beyond National Jurisdiction was turned into a Preparatory Committee for the Law of the Sea Conference, Wooster, as President of SCOR, sought to involve the organization in the preparations for the Conference. During the tenth General Meeting of SCOR, he made a proposal "to establish a working group to evaluate the need for freedom of scientific research in the ocean and the obligations implied by that freedom and to examine the consequences for scientific research of the various alternative ocean regimes." While it was agreed that the Executive Committee should solicit the views of a large number of scientists and national committees on this issue, there was strong feeling that "by becoming involved in such a controversial problem, SCOR could weaken its credibility as a scientific body" (SCOR, 1970).

The responses to this solicitation reflected this concern among most National Committees. Only the U.S. committee was unequivocally in favor of active involvement. Those of Australia, France, the United Kingdom, and the USSR were equally unequivocally opposed (SCOR, 1971). Consequently, SCOR did not become actively involved as an organization. As an alternative, a statement on the characteristics of marine scientific research, originally drafted by Wooster, was circulated to all members with the intention that National Committees would pass this on to their government's Law of the Sea delegations. During the General Meeting in 1972, Professor Ilmo Hela of Finland proposed that SCOR offer its services to the UN Secretariat for the Conference in an attempt to provide scientific advice on issues affecting marine science (SCOR, 1972: 11). This, too, was not adopted.

The formal involvement of ICSU/SCOR at the Law of the Sea Conference has been limited to general statements of Professor H. Postma, the President at the time, to the Third Committee during the Spring 1975 Sessions at Geneva, an official SCOR/ICSU statement transmitted to the Conference. These actions had no impact on the official positions of governments.

Why did this happen when, ostensibly, the trend in the negotiations was so adverse to the interests of oceanographers, in the sense that a large amount of new restrictions were being imposed? The answer to this question is in two parts. The first is that for the European scientists especially, involvement in such an openly controversial, political negotiation appeared to be a violation of their perceived roles as scientists. (The Americans had no such inhibitions.) But, more importantly, the answer would seem to be in governmental perceptions of national interests on this issue and the nature of the relationship between governments in most advanced maritime countries and their oceanographers. This was inadequately appreciated by U.S. scientists until the New York Sessions of the Law of the Sea Conference in Spring 1976.

The point is that only the USSR and the United States have global programs in the ocean. However, Japan, the United Kingdom, the Federal Republic of Germany, and the German Democratic Republic have significant distant-water research programs. While scientists of other nationalities certainly have global interests in the ocean, their programs are more locally or regionally based. The consequences of increasing restrictions on scientific research within 200 miles of the coasts of different countries, therefore, were major problems only for a very few countries. Moreover, since the fundamental tradeoff at the Law of the Sea Conference was posed as a choice between extending coastal state control over resources versus protecting the navigational and research rights of others, governments like the United Kingdom, Canada, Norway, and Australia, which would normally support the interests of science, chose to extend control over resources. As a result, they kept their scientists on very tight rein at the Conference, and the United States found itself for most of the time fighting off the Group of 77 with only a few countries as allies. The final defection came in the Spring of 1966, when the USSR suddenly switched tactics and supported the Group of 77. The choice appeared to have been made because the USSR was much more concerned to exclude or control U.S. research in the Soviet Economic Zone than in being excluded itself in other parts of the world.

The two crucial conditions for success in employing a transnational strategy, therefore, never existed. SCOR did not have a unified point of view, and scientists in most countries were not allowed access to governmental centers of decision-making where official perceptions of national interest potentially diverged from those held by the scientists. The level of control where disagreement was high, as in the case of Canadian scientists, was dependence on the government for funds to engage in scientific research.

CONCLUSIONS

Both COSPAR and SCOR have had significant impacts on the conduct of certain kinds of space and marine science within developed countries. Both have also facilitated increased allocations for scientific investigations from their governments. In each case, however, the impact of SCOR appears to be considerably larger than that of COSPAR. COSPAR is constrained by the fact that it does not have a large constituency and is overly dependent upon and sensitive to changing relations between the superpowers. Until the late 1960s, COSPAR was important to the U.S. government because it provided a flexible framework for dealing with the Soviet Union. As the climate improved between them, the need for COSPAR decreased significantly. The demand for COSPAR's advice is not great, and it therefore cannot build an effective constituency independently of the United States and the USSR. SCOR, on the other hand, is not dependent primarily on the superpowers, since the pool of players with advanced marine capabilities is much larger. The viability of SCOR derives from the actual and potential demand for its services.

At the same time, it should be noted that SCOR's fortunes are responsive also to changes in its internal and external environment. Internally, there appears to be a major problem in recruiting individuals to succeed those currently responsible for steering the organization. Since the administrative leadership is small, changes in the individual ability, preferences, and style of the President appear to have very significant impact on the vitality of the organization.

The essential point to be made, therefore, about the *long-term* political significance of COSPAR and SCOR is that it is episodic and perhaps even haphazard, because there are several important variables, control over which is problematic. We refer to: (a) the quality of political leadership, (b) funding, (c) the viability of the intergovernmental framework to which the INGO is attached, (d) specific changes in the external environment, (e) the emptiness of global scientific planning; and (f) culturally determined differences in role perception.

Though not all of these variables are to be given equal weight, there are lessons here for all ICSU Special or Scientific Committees whose work touches upon resources which governments want to see divided. The situation presents something of a Hobson's choice for the ICSU leadership because science, technology and resource utilization, and distribution questions are currently at the top of the international agenda. If ICSU committees refrain from giving policy advice of whatever kind to their intergovernmental clients, these organizations will simply create their own science advisory bodies, thereby depriving the ICSU units of useful sources of funds. On the other hand, because these issues are so controversial, active involvement in the policy game also entails high risk, and success in navigating such troubled waters is clearly a matter of skill.

NOTES

1. The only exception to this criticism is Keohane and Nye (1971).

2. For some program description, see ICSU (1969); also see Baker (1972). This pamphlet, written by the Executive Secretary, is an effective introduction to the ICSU for those unfamiliar with the organization.

3. Recently, the scope of COSPAR's concerns has been expanded to include scientific research with balloons.

4. See, for instance, the proposal of the U.S. Department of Defense to orbit a space reflector to illuminate the jungles of Vietnam at night; Clark (1966).

5. This account was previously published in Miles (1971). It is included here for purposes of comparison with the attitude of SCOR to the Third UN Conference on the Law of the Sea.

6. Prior to the creation of SCOR, interdisciplinary discussions on oceanography occurred episodically with the Joint Commission on Oceanography formed between the International Union of Geodesy and Geophysics and the International Union of Biological Sciences.

7. At the same time, the Office of Oceanography was created within the UNESCO Science Sector. But this was a purely internal, administrative unit without the significant potential of IACOMS.

8. For a brief analysis of the IIOE experience, see Iselin (1959), Deacon (1957), Knauss (1961), and Humphrey (1963). The type of government support required went far beyond direct financial contributions. For instance, governmental agreements were required for the use of deep sea buoys, assignment of frequencies for vessels and instruments, access to territorial seas, and the like.

9. Personal communication to the author, May 11, 1973.

REFERENCES

ANGELL, R. (1969) Peace on the March. New York: Litton.

BAKER, F.W.G. (1972) The International Council of Scientific Unions: A Brief Survey. Rome: ICSU Secretariat.

British National Committee for Oceanic Research, The Royal Society (1963) Need for Expansion in Marine Science. NOR/11: 1.

CLARK, E. (1966) "Defense study of space mission stirs fears of blow to astronomy." New York Times (December 29): 8.

Committee on Space Research [COSPAR] (1971) "Fourteenth Plenary Meeting, including symposia and open meetings of Working Groups, June 17-July 2, 1971." First Circular, December 1970: 2, 1.3.

--- (1969) Information Bulletin 4 (March): 8-9.

--- (1966) Information Bulletin 31 (April): 4-11.

--- (1965) Information Bulletin 22 (February): 88.

--- (1964) Information Bulletin 20 (November): 20.

--- (1963) Information Bulletin 13 (January): 14-16.

--- (1961) Information Bulletin 5 (July): 10; (November): 20-21.

--- (1960) Information Bulletin 2 (June): 4-5.

DEACON, G. (1957) "International cooperation in marine research." Nature 180 (November): 894-895.

FRUTKIN, A. (1965) Space and the International Cooperation Year. Washington, DC: Government Printing Office.

HUMPHREY, G. (1963) "The international Indian Ocean expedition." Hemisphere 7, 4: 4-7.

International Council of Scientific Unions (1969) "Note on ICSU activities in the field of scientific and technological cooperation." Prepared for the European Economic Commission Meeting of Governmental Experts on Technology. 6: Add. 8, 1-8.

International Telecommunication Union (ITU) (1963a) Extraordinary Administrative Radio Conference (EARC). Doc. 8-E, Annex; Doc. 46-E, Annex.

--- (1963b) Extraordinary Administrative Radio Conference (EARC). Doc. 75-E.

ISELIN, C. (1959) Report of the Indian Ocean Working Group to SCOR, C. Iselin, chairman. (mimeo).

JOHNSON, J. (1964) "Pollution and contamination in space," pp. 37-50 in M. Cohen (ed.) Law and Politics in Space. Montreal: McGill Univ. Press.

KEOHANE, R. and J. NYE, Jr. [eds.] (1971) Transnational Relations and World Politics. Cambridge, MA: Harvard Univ. Press.

KNAUSS, J. (1961) "The international Indian Ocean expedition." Science 134 (November): 1674-1676.

LADOR-LEDERER, J. J. (1963) International Non-Governmental Organizations. Leyden: A. W. Sythoff.

MILES, E. (1971) "Transnationalism in space: inner and outer," pp. 602-625 in R. Keohane and J. Nye, Jr. (eds.) Transnational Relations and World Politics. Cambridge, MA: Harvard Univ. Press.

National Aeronautics and Space Administration, U.S. [NASA] (1975) Objectives and Models of the Planetary Quarantine Program. NASA SP-344. Washington, DC: Author.

ODISHAW, H. [ed.] (1962) The Challenge of Space. Chicago: Univ. of Chicago Press.

ROY, M. (1969) "Benefits of membership in COSPAR." COSPAR Information Bull. 48: 5-10.

SAGAN, C., E. C. LEVINTHAL, and J. LEDERBERG (1968) "Contamination of Mars." Science 159 (March): 1191.

Scientific Committee on Oceanic Research (SCOR) (1972) SCOR Proceedings 8 (April): 11.

--- (1971) SCOR Proceedings 7 (September): 25-34.

--- (1970) "Necessary conditions for the effective conduct of ocean research." Circular Letter to Members and National Committees, 529D (December).

--- (1963) "Letter of A. Maxwell to G. Humphrey, President." Sixth Meeting of SCOR (April 12).

--- (1962a) Report of the Meeting of the SCOR Executive at London with the UNESCO Office of Oceanography, April 16-18. (mimeo)

--- (1962b) Report of the Secretary of SCOR to ICSU, July 6. (mimeo).

SKJELSBAEK, K. (1971) "The growth of international nongovernmental organizations in the twentieth century." Int. Organization (special issue edited by R. Keohane and J. Nye, Jr.) 25, 3 (Summer): 420-442.

SKOLNIKOFF, E. B. (1967) Science, Technology, and American Foreign Policy. Cambridge, MA: MIT Press.

SMOKER, P. (1965) "Preliminary empirical study of an international integrative subsystem." Int. Associations 11 (November): 638-645.

SNEATH, P.H.A. [ed.] (1968) COSPAR Technical Manual 14: Committee on Space Research.

Special Committee on Oceanic Research (1957) Report to ICSU on First Meeting, August 28-30. (mimeo)

WENK, E. (1960) Radio Frequency Control in Space Telecommunications. Prepared for the use of the Committee on Aeronautical and Space Sciences, U.S. Senate, 86th Congress, Second Session. Washington, DC: Government Printing Office.

WHITE, L. (1951) International Non-Governmental Organizations. New Brunswick, NJ: Rutgers Univ. Press.

WOOSTER, W. S. (1973) "Interactions between intergovernmental and scientific organizations in marine affairs." Int. Organization 27, 1: 103-114.

--- (1969) "Should there be an international union of marine sciences?" (unpublished)

8

STOCKHOLM AND VANCOUVER: THE ROLE OF ISPAs AT UN CONFERENCES

Anne Thompson Feraru

The United Nations Conference on the Human Environment held at Stockholm in June 1972 was the first of the "megaconferences" on major world problems which have been held in recent years under UN sponsorship. Others include the World Population Conference at Bucharest and the World Food Conference at Rome in 1974, the International Women's Conference at Mexico City in 1975, the UN Conference on Human Settlements (Habitat) at Vancouver in June 1976, and the Conference on the Law of the Sea which has met intermittently during 1974-1976 at Caracas, Geneva, and New York. Two additional conferences on water and desertification were held in 1977, and a 1979 conference was convened on the use of science and technology for development.

This remarkable series of international conferences on "functional" issues of world concern, "low" politics as contrasted with the "high" politics which have been the traditional focus of intergovernmental meetings, has attracted the participation of a large number of nongovernmental organizations, both international and national or subnational in membership. NGOs (or INGOs) have been involved in the conference preparatory activities of governments and international organizations; they have sent official observers to the conferences and have taken part in nonofficial gatherings held concurrently with the governmental meetings; and have continued to be involved in the implementation of conference recommendations by international agencies and by governments.

This chapter concerns NGO activities relating to two of the problem-focused conferences, the 1972 Stockholm Conference (UNCHE), which was the first of the series, and the 1976 Human Settlements Conference, which was conceived as a follow-up to UNCHE in that its agenda included consideration of the impact of the man-made environment on the natural environment. The NGOs discussed here are limited to *international* nongovernmental organizations (thus excluding the national and subnational groups whose members took part in conference-related activities at Stockholm and Vancouver) and further limited to INGOs in scientific and professional fields which had accredited observers at one or the other of the two intergovernmental conferences.

In successive sections, those scientific and professional INGOs involved in Stockholm and Habitat are identified, INGO perceptions of their objectives in attending the conferences are analyzed, the points of contact between them and international and government officials prior to and at the conferences are described (and in the case of UNCHE, their relations with the UN Environment Programme and the Governing Council since Stockholm), and some speculations as to the conditions under which INGOs can exert influence on the decision-making processes at United Nations conferences are drawn.

Information for this chapter, in addition to UN documents and other published materials, is drawn from mail surveys of INGO participants at UNCHE and Habitat and from direct observation at the Vancouver Conference.

ISPA PARTICIPANTS AT UN CONFERENCES

Which ISPAs Were Involved in the Stockholm and Vancouver Conferences?

Of the 2470 international nongovernmental organizations catalogued in the 1974 *Yearbook of International Organizations* (Union of International Associations, 1974: 533), 178 had one or more official observers at the Stockholm Conference and 145 at Habitat, representing organizations from all of the fields of interest according to which the Union of International Associations classifies INGOs. Our concern here is with the subset of INGOs identified by Evan (1974: 605) as falling in the category of "scientific" and "professional" associations (ISPAs). Specifically, these are organizations classified by UIA in the following seven categories: social sciences; law, administration; professions, employers; economics, finance;

technology; science; health, medicine. Of the 901 organizations listed in the *Yearbook* in these seven categories, 73 had observer status at Stockholm and 36 at Vancouver. The totals for each category are listed in Table I, and the organizations are identified in Appendix A.

Analysis of Appendix A shows that 24 of the ISPAs at UNCHE were also represented at Habitat, while 49 were at Stockholm only and 12 at Vancouver only, a total of 85 different organizations at either or both conferences. The "repeaters" are identified by asterisks in Appendix A.

Some of those same organizations which attended both conferences are also among those with the largest number of accredited observers, as shown in Appendix A. Most of the ISPAs were represented by one or two persons, but at each conference, there were six organizations with five or more observers. At Stockholm, these six organizations (Friends of the Earth, ICSU, IUCN, Sierra Club, World Wildlife Fund, World Energy Council) accounted for 29 percent of the 154 observers from scientific and professional associations; at Habitat, the six organizations with five or more observers (FOE, Sierra Club, Commonwealth Human Ecology Council, International Union of Architects, International Union of Local Authorities, World Society for Ekistics) accounted for 60 percent of the 133 ISPA observers.

ISPA OBJECTIVES AT UN CONFERENCES

What Were ISPA Objectives in Sending Representatives to the Conferences?

Before describing the activities of ISPAs and other INGOs at the conferences and identifying their points of contact with government and Secretariat officials, it might be useful to consider *why* they sent representatives to the conferences. It seems plausible to assume that the use they make of the opportunities for interaction afforded at intergovernmental conferences would depend to a considerable extent on the objectives which the INGOs themselves had in attending the conferences.

Some information on those objectives is available from responses to a mail survey sent in January 1973 to all INGOs with observers at Stockholm and a similar survey sent in August 1976 to the international scientific and professional associations at Habitat. Responses to the Stockholm survey were received from 43 percent (81) of the 178 INGOs in all categories,[1] and from 59 percent (43) of the 73 ISPAs at Stockholm; replies to the Habitat survey are less complete, with only 28 percent (10) of the 36 ISPAs responding.

The responses indicate that some, but not all, of the INGOs conceive of themselves as "interest groups" whose objectives in attending the confer-

TABLE 1: International Scientific and Professional Associations, By Category: Total Number of Organizations, Numbers with Observers at UNCHE and Habitat

Category	*Total Number of Organizations**	*Observers at UNCHE, 1972***	*Observers at Habitat, 1976****
Social Sciences	104	9	7
Law, Administration	58	11	3
Professions, Employers	119	12	7
Economics, Finance	47	2	1
Technology	133	13	5
Science	184	15	8
Health, Medicine	256	11	5
TOTAL	901	73	36

SOURCES: *Union of International Associations, 1974.
**UN Document A/CONF. 48/INF.6.
***Conference Registration List—Delegates, June 6, 1976, pp. 51-72.

ences were to articulate the views of their organizations' members and to exert pressure upon the official international decision makers. The "interest group" objectives mentioned in the survey responses fall into two related but somewhat distinguishable types, which I have labeled "lobbying" and "representation," "lobbying" objectives being to urge consideration or adoption of some particular principle or program, and "representation" objectives referring to less specific aims of representing or expressing the general concern or interest of the organization in the subject of the conference.

The "political" objectives of lobbying and representation were identified by over half of all INGOs responding to the survey concerning Stockholm, but by only one-fifth of the responding ISPAs at Stockholm. Further evidence of the relatively low level of ISPA interest in making their views known is provided by their answers to questions asking, "Did your organization—at or prior to the Conference—have occasion to communicate views on the environment to officials of the UN Secretariat? to officials of governments?" Of the 43 organizations, only 14 reported communicating with governments prior to, and 10 at, Stockholm; 20 communicated with the Secretariat prior to, and 7 at, the Conference. The very fragmentary information on ISPAs at Habitat shows five of the ten responding organizations mentioning the interest articulation objectives of lobbying and representation; seven communicated views to governments at or prior to Habitat and six communicated with Secretariat officials.

Other reasons given for sending observers to the conferences were *to get information* about what happened there for transmission to the organization's members or to the wider public, and *to give expert assistance* in the form of advice or information on some aspect of the conference agenda. More than half of the ISPAs mentioned getting information as an objective in going to Stockholm, and over one-quarter of them identified providing expert advice as an objective.

Table 2 summarizes responses to the surveys, and the responding organizations are identified in Appendix A.

ISPA CONTACT WITH GOVERNMENT AND IGO OFFICIALS

What Points of Contact with Government and IGO Officials?

INGOs have a number of opportunities to interact with government and international Secretariat officials at the conferences and in the pre- and postconference periods. The major points of contact are listed below, followed by discussion of the use of those contact points by the ISPAs at UNCHE and Habitat.

TABLE 2: Responses to Survey Question, "What were your organization's objectives in sending representatives to the conference?"

	All INGOs at UNCHE (N = 81)	*ISPAs at UNCHE (N = 43)*	*ISPAs at Habitat (N = 10)*
"Lobbying"—to urge consideration or adoption of some particular principle or program	20	6	2
"Representation"—to communicate the interests, concerns, viewpoint of members	23	8	4
"Information"—to get information, observe, find out what's going on	38	23	1
"Expertise"—to provide expert advice or information	13	12	2
Other	5	2	3
No Response	2	2	2

NOTE: More than one answer could be given to the question; therefore, the number of responses may exceed the total number of organizations that returned the survey questionnaire.

Before the conference:

(1) attendance as observers at Preparatory Committee meetings
(2) participation on Intergovernmental Working Groups drafting conference documents
(3) submission of reports to Conference secretariat
(4) participation in "unofficial" international conferences in preparation for the official intergovernmental conference
(5) membership on national advisory bodies
(6) communication of views to governments by national sections of international associations

At the conference:

(1) attendance as observers at plenary and committee sessions
(2) submission of written statements which are circulated to delegates as part of the official conference documentation
(3) addressing the plenary meeting, either as representative of a particular INGO or on behalf of a group of INGOs
(4) "lobbying" in corridors at official meetings
(5) attendance at briefings by national delegations and by the UN Office of Public Information
(6) attendance at social events, such as receptions
(7) participation in INGO "parallel" conference, held simultaneously with intergovernmental conference
(8) participation in other events, such as press conferences, marches, rallies
(9) publication of INGO newspaper during conference

After the conference:

(1) attendance as observers at ECOSOC and General Assembly during consideration of conference recommendations
(2) submission of written statements to ECOSOC at time of consideration of recommendations adopted at the conference
(3) attendance as observers at meetings of UN bodies created as a result of the conference
(4) provision of services to postconference UN bodies
(5) participation in postconference INGO group with a continuing interest in the subject of the conference

Stockholm: ISPA Activities Before the UN Conference

Scientific and professional INGOs and individuals associated with them were actively involved in preparations for the UN Conference on the Human Environment at the request of the Preparatory Committee, the Conference secretariat, various specialized agencies, and governments preparing delegations' positions, and on their own initiative.

In the work of the 27-country Preparatory Committee, observers from eight professional INGOs attended one or more of the four sessions held in the period 1970-1972.[2] Experts associated with ISPAs also participated as members of the Preparatory Committee's intergovernmental working groups; the International Union for the Conservation of Nature and Natural Resources (IUCN) was represented on the Conservation Group, the International Society of Soil Science on the Soils Group, and the International Council of Scientific Union's Scientific Committee on Problems of the Environment (SCOPE) on the groups on monitoring and the declaration on the human environment. ICSU, IUCN, and other nongovernmental organizations were invited to prepare basic papers on subjects related to the Conference agenda; consultants employed by the secretariat used the basic papers in drafting position papers which, in turn, were circulated among a panel of experts for comments.[3]

One of the most publicized preparatory documents was the report on the state of the human environment commissioned by the Secretary-General for UNCHE, Maurice Strong. Intended for a lay audience and published prior to the meeting as *Only One Earth* (Ward and Dubos, 1972), this report was prepared by the biologist Rene Dubos and the social scientist Barbara Ward. The draft of *Only One Earth* was circulated for comment among a 152-member Committee of Corresponding Consultants in 58 countries; the Committee included natural, social, and political scientists, more than two-thirds of them experts in scientific and technical fields, and 28 affiliated with scientific and professional INGOs.

In addition to the Preparatory Committee sessions and the preparation of documentation for UNCHE, a number of conferences were organized by, or in cooperation with, the Conference secretariat in 1971. Experts, government and UN officials participated in a conference on International Organization and the Human Environment, held in May 1971 at Rensselaerville, New York, under the auspices of the Institute on Man and Science and the Aspen Institute for Humanistic Studies, and in the Founex meeting on Environment and Development in June 1971. ICSU's SCOPE convened a working party of scientists at Canberra in August-September 1971, and UN Economic Commissions convened four regional seminars in preparation for Stockholm. At these meetings, multinational groups of scientists and other experts exchanged views with each other and with specialists in UN and government service.

Expert views were also solicited by governments in their preparations for Stockholm (see National Academy of Sciences, 1972), and reports and recommendations on international environmental action were prepared by a number of private organizations including INGOs, some of which were included in the Conference bibliography.[4]

Stockholm: ISPA Activities At the UN Conference

At the Stockholm Conference, 73 of the 178 international nongovernmental organizations which sent observers were organizations in professional and scientific fields. Observers attended the public meetings of the Conference and its main committees, and at least one representative of a professional INGO, Margaret Mead, anthropologist and observer for the World Society for Ekistics, was accorded the privilege of making an oral statement in the plenary meeting on behalf of individuals affiliated with INGOs (UN Centre for Economic and Social Information, 1972: 17-18).

In addition to observing at the official conference sessions, INGOs—both national and international—were able to keep informed of developments through daily briefings held by the UN Office of Public Information, and had opportunities to interact with national delegates at receptions and other social events. The Environment Forum, the unofficial conference for INGOs held in Stockholm simultaneously with UNCHE, provided another setting in which some 200 groups, most of them organizations without UN observer status, could express their views through lectures, exhibits, parades, and other activities which presumably were intended to reach the governmental representatives, as well as the attentive public throughout the world.

Another vehicle for INGO-government communication during the Conference was the *ECO*, a daily newspaper produced jointly by the Friends of the Earth and the magazine, *The Ecologist*. The *ECO* contained editorials, conference news summaries and analyses, and was distributed free to government and INGO delegates. It was apparently widely read by government representatives at UNCHE, in part because the other news coverage easily available in Stockholm was in the Swedish language; the English-language *ECO* filled a news "gap" for many delegates, and communicated INGO views on conference issues as they came up each day.

Stockholm: ISPA Activities After the UN Conference

In the post-Stockholm period, INGOs attended conferences bringing together the wide range of international and national nongovernmental organizations interested in the world environment. Two were held in October 1972, one in Geneva and one in New York at the time of ECOSOC and General Assembly consideration of the Stockholm recommendations. Since the creation of the UN Environmental Programme with headquarters in Nairobi, INGOs have formed the International Assembly of Non-Governmental Organizations concerned with the environment, which has held conferences at the time of sessions of UNEP Governing Council, and they have established the Environment Liaison Centre in Nairobi to coordinate activities of environmentally concerned INGOs and to facilitate INGO contacts with the UNEP secretariat and Governing Council.

Several ISPAs, most notably ICSU's SCOPE and IUCN, have close working relationships with the UNEP secretariat, and perform contract services for UNEP programs, such as Earthwatch, a program for research and evaluation of international environmental problems, and the International Referral System being established to coordinate exchange of information among a variety of producers and users of data on the environment.

Vancouver: ISPA Activities Before the UN Conference

The UN Conference on Human Settlements, as a sequel to UNCHE, attracted the participation of a number of those organizations which had been involved in Stockholm. Preparatory activities, like those for UNCHE, included attendance as observers at the two meetings of the Preparatory Committee, submission of reports and statements to the Habitat Secretariat headed by Secretary-General Enrique Penalosa and to national governments preparing their positions, the participation of individual architects, planners, environmentalists and other professionals at a symposium convened by the secretariat at Dubrovnik in May 1975, and the participation of INGOs interested in Habitat at a meeting held in Geneva in Fall 1975.

In addition, representatives of international and national private agencies comprised the NGO Committee for Habitat which, in cooperation with a private Canadian organization (the Association in Canada Serving Organizations for Human Settlements) made preparations for Habitat Forum, the parallel conference of organizations and individuals interested in the agenda of the intergovernmental conference. On the 14-member NGO Committee for Habitat were four persons affiliated with international professional associations: Zena Daysh of the Commonwealth Human Ecology Council (CHEC), Dr. S. El Hamamsi of the World Society for Ekistics, Jean-Louis LaLonde of the International Union of Architects, and J. G. van Putten of the International Union of Local Authorities.

Vancouver: ISPA Activities At the UN Conference

At Vancouver, 36 of the 145 organizations which were accorded official observer status at Habitat were INGOs in professional and scientific fields; most of them also participated in activities at the Habitat Forum. At the governmental Conference, representatives of INGOs with official status were permitted to attend the plenary sessions and meetings of the three committees deliberating on the draft documents before the Conference.

During the open meetings of the committees, INGO observers sat at the back of the room, behind the desks of the official participants; other than the infrequently observed tactic of moving into the delegates' seating area to deliver a note or confer briefly, the contacts during the official sessions were limited to conversations between observers and governments' representatives in the lounge areas and to a few written statements by INGOs which were circulated as part of the official Conference documentation. However, on two occasions, an exception was made to the usual practice of restricting participation in committee debates to official delegates. Committee I permitted a spokesman for a group of 11 INGOs to make an oral statement urging that the resolutions on the postconference institutional arrangements include a statement recognizing the need for INGO involvement in such activities; among the groups joining in the statement were such ISPAs as IUCN, CHEC, Sierra Club, Union of International Associations, and the International Union of Architects. The second occasion was in Committee III, which heard an oral statement by the representative of the International Federation of Women in Legal Careers.

At the plenary sessions, J. G. van Putten, Secretary-General of IULA and chairman of the NGO Committee for Habitat, spoke on two occasions to present to the delegates statements on behalf of groups and individuals participating in Habitat Forum, and Robert Allen of IUCN was one of the dozen other INGO representatives who were permitted to make brief addresses to the plenary.

Other occasions for contact with government delegations were afforded by the daily briefings held by some country delegations at their headquarters in downtown Vancouver, and at the Habitat Forum site to which the United States and other government representatives came for special meetings with members of INGOs and unaffiliated individuals from their own countries; some government delegates also went to the Forum to speak or listen at panels arranged by INGOs on topics related to the conference agenda, such as national community development programs. Some of the delegations, such as the United States, included among their advisers persons associated with national NGOs.

Receptions and social gatherings put on by national delegations, the host city and others provided occasions for contacts; FOE took the initiative of renting a house in Vancouver where it hosted a series of "salons," informal dinners to which certain delegates and others attending the conference were invited.

In addition to these direct contacts between the participants in the UN Conference and the Forum, nongovernmental organizations and individuals used indirect means for communicating views to the official decision makers. These included press conferences (for example, by Friends of the Earth to publicize their support for a moratorium on nuclear power plant construction) and public demonstrations (such as the World Water Walk to the Forum to press for conference adoption of a resolution urging governments to provide clean water for everyone by 1990, and an anti-nuclear demonstration held in downtown Vancouver) which were covered by the media. Just prior to the conference, a number of internationally known individuals (among them Barbara Ward, Margaret Mead, Maurice Strong) participated in the Vancouver Symposium and produced a declaration on major issues before the Conference, as well as a second statement during the Conference.

The Forum daily newspaper, Jericho (named for the site of the Forum at Jericho Beach across the bay from downtown Vancouver where the UN Conference was held), did not seem to have an impact on delegates comparable to the Stockholm *ECO,* at least in part because Vancouver newspapers, unlike Stockholm's, are in English, so delegates did not have to rely on the Forum newspaper for coverage of conference events. *Jericho* seemed to serve primarily as a means of inter-NGO communication about events going on at the Forum and at the UN Conference

SOME OBSERVATIONS ON INGOs AT HABITAT

While opportunities for direct and indirect contacts between governments and NGOs existed, effective political action by nongovernmental organizations was impeded by several factors. First, the physical locations of the official conference and the Forum–as at Stockholm–were separated by some miles. The intergovernmental conference was held in various hotels and a theatre in downtown Vancouver, while the Forum was held in a converted air base in another part of the city; the two sites were linked by a special bus service. INGOs also had meeting rooms and a lounge, equipped with TV to monitor the official meetings, at a downtown hotel where the Canadian Habitat Lobby for INGOs maintained an office.

Second, information about what was going on at the official conference was not easily available to persons at the Forum. Many of them did not have copies of the basic conference documents nor were copies of the draft resolutions under consideration in the three committees made widely available to NGOs; the professional INGO representatives, with official UN observer status, did have access to the UN documents office in downtown Vancouver, but this privilege was not extended to most of the groups at the Forum who were "nonstatus" INGOs. *Jericho* and Vancouver newspapers provided some coverage of conference developments, but not in detailed enough terms to be useful to INGOs interested in modifying specific clauses of the resolutions being considered by the conference. The daily briefings by OPI to Forum participants came on the morning following each day's conference sessions, rather than at the end of each day when INGOs might have been able to use the evening to prepare for action on the following day. Graham Searle of the Habitat Lobby for INGOs instituted evening briefing sessions at which reports were given by monitors who attended the meetings of each of the three committees and the plenary; Searle's group–less constrained than OPI personnel to observe international civil service standards of neutrality–attempted to direct INGO attention to the political significance of various moves within the committees. But the general impression was that, although facilities for NGOs both downtown and at Jericho Beach were better organized than at Stockholm, many of the NGO representatives at the Forum had only fragmentary knowledge of what was going on at the UN Conference and little idea of what use to make of the information they did have.

The lack of information is related to another factor which limited Forum impact on the UN Conference–that is, the "mixed bag" of persons at the Forum. Lumped together under the "NGO" label were a number of individuals unaffiliated with any organized group, a number of ad hoc coalitions of special interests such as the antinuclear power group, some *subnational* groups such as Canadian Indians who tried to use the international conference as a place to influence their own government's policies, *national* NGOs, as well as *international* NGOs with international memberships and international objectives. Expertise in how to influence official delegates at an international conference seemed to me to be largely limited to persons in the last category, many of whom are veterans in IGO-INGO relationships and "alumni" of previous international conferences at Stockholm, Bucharest, and Rome. The diversity in types of organizations, objectives, and level of political sophistication made it very difficult for the Forum to arrive at consensus on the joint NGO statements delivered to the UN Conference. Dissatisfaction among Forum participants with the first statement, prepared from position papers submitted to the NGO Committee prior to Habitat, led to demands for the second statement; prolonged discussions among NGOs on the content of that statement delayed its presentation to the conferences until the last day, when it was too late to have an impact on the conference resolutions. My impression is that many of the Forum participants felt frustration at their apparent inability to affect the outcome of the conference, frustration which was directed both at the NGO leadership and at the "conservatism" of government representatives.

While the separate locations of UN Conference and Forum, the lack of information about what was going on at the UN Conference, and the lack of expertise in how to reach government delegates seem to have adversely affected the impact of many of the NGOs, these observations are less applicable to at least some of the scientific and professional INGOs which are the subject of this chapter. As INGOs with consultative status, they had access to the meeting rooms of the UN Conference and to the documents office maintained by the UN for the use of official delegates. Some of those organizations (notably IUCN, IULA, CHEC, Sierra Club, and Friends of the Earth) were represented by people with prior experience at intergovernmental conferences who "knew the ropes" of how to get information, how to identify and approach key delegates directly, and how to form ad hoc coalitions of NGOs on particular items of common interest. They seemed to concentrate their efforts to influence governments through activities downtown, rather than through the Forum, which functioned more as a place for communicating information and ideas about human settlements among NGOs and to the public (and, some participants suspected, to keep NGOs occupied elsewhere while the government delegates were deliberating at the official meetings downtown).

The extent to which these ISPA efforts succeeded in influencing the outcomes of the Conference is impossible to characterize with any degree of assurance. Detecting and measuring the relative influence of various participants in decision-making processes is a difficult enterprise at any level of political analysis and particularly so at the international level; however, some tentative, highly speculative observations on the influence of ISPAs can be offered as a stimulus to further research and analysis.

POLITICAL IMPACT OF ISPAs AT UN CONFERENCES

In assessing the political impact of INGOs in general and ISPAs in particular on international decision-making at international conferences such as UNCHE and Habitat, two polar views are likely to be held, one which overestimates the impotence of nongovernmental organizations and one which overestimates their influence. According to the first view, governments are the decision makers and, although some governments' positions on some issues may be responsive to the expressed interests of some nongovernmental groups within their own countries, governments generally are likely to be unaffected by the activities of INGOs. Evidence to support this view is found in the very limited extent to which the draft resolutions under consideration at intergovernmental conferences are modified in response to pressures by INGOs. If influence is measured by changes effected in resolutions adopted at the conference, one might conclude that INGO influence is negligible.

The other polar view, in contrast, sees the presence of several hundred NGO observers at intergovernmental conferences, and the presence of perhaps several thousand additional interested persons representing a variety of nongovernmental groups at "parallel" conferences, as evidence that governments' monopoly over international decision-making has been successfully challenged. In this view, the "world-oriented" nongovernmental forces present at international conferences are increasingly able to modify the parochial "national interest" orientation of governmental representatives and to influence governments to adopt positions which are advocated by nongovernmental interests.

Neither of these views seems to me an accurate description, in part because the first view defines "influence" too narrowly, because the second exaggerates both the homogeneity among governmental representatives on the one hand and INGOs on the other and the antagonism between them, and because both views focus exclusively on the few days of the conference as *the* arena of decision-making and the resolutions adopted as the significant "output."

My observations, which are offered as propositions to be verified rather than as firm conclusions, are:

First, the opportunities for effective INGO influence on national and international decision-making are less *at* intergovernmental conferences than *before* them, through contacts with IGO and government officials. International conferences are short in duration, and delegates are busy attending plenary and committee meetings, delegation meetings, and caucusing group meetings. Because time is short, discussion centers on draft resolutions and declarations prepared by the Secretariat and reviewed by the intergovernmental preparatory committee; the likelihood of major substantive changes being made at the Conference is not great. Therefore, it would seem that the preparatory phase, when draft documents are being produced which are the basis for discussion at the conference, would offer greater opportunity—especially for ISPAs—to influence the outcomes of intergovernmental deliberations than does the brief and hectic period of the conference itself.

Second, opportunities for effective INGO influence are afforded in the *post*conference period, when resolutions adopted at the conference are being implemented by governments and through the establishment of international institutions (such as UNEP) and the development of international programs (such as Earthwatch). In the postconference phase, the international Secretariat charged with program development and implementation is likely to welcome the technical and scientific expertise of ISPAs, as well as the capacity of INGOs to mobilize support for such programs among the attentive publics in the member countries.

Third, evidence of the influence is INGOs is to be found not simply (and perhaps least) in modifications of and additions to formal resolutions adopted at intergovernmental conferences, but in formal and informal decisions taken within governments, within the international secretariat, within the various organs of the UN and other international agencies, during the whole stream of continuous decision-making, of which international "problem" conferences are only one phase.

More specifically, INGOs may be influential in "setting the agenda" for governments and IGOs, by urging international attention to problems which—until recently—have been regarded as outside the realm of traditional international politics. Once governments are prepared to discuss such problems in an international forum, INGOs may exert further influence in framing the options open to governments and IGOs in dealing with the problem, and subsequently in providing advice in program development and services in program implementation. Incremental decisions taken at each step along the way are likely to afford opportunities for INGO involvement, at least for those with the reputation, the human and financial resources, the political insight to know who is deciding what and when, and the access to those "official" decision makers.

Some ISPAs—ICSU and IUCN most notably—do seem to possess such attributes. However, ISPAs (and INGOs generally) do not seem to function as a single group, although ad hoc coalitions of some of them may form on occasion, such as at Vancouver when a number of INGOs combined to present the case for formal recognition of INGO involvement in the post-Habitat institutional arrangements. The great diversity of interests and resources among the INGOs reduces their capacity to act collectively, and—for the researcher—suggests that looking for generalizations about the influence of INGOs as a class of actors will provide fewer pay-offs at present than narrower-gauge descriptions of particular organizations' involvement in specific decisions.

These observations lead to a conclusion concerning the context of international policy processes in which INGOs participate. That context is more complex than depicted by the conventional view which identifies governments as the authoritative decision makers and IGOs as institutions through which states, each acting as a unit, make and implement policy in the international system. A different approach to international policy processes, suggested by Keohane and Nye, offers another way of looking at the political role of transnational actors such as INGOs (Keohane and Nye, 1974; Keohane, 1975; Ruggie, 1972). The approach Keohane and Nye (1974: 41) suggest is to "relax the assumption that states are the only actors, and . . . the assumption that states act as units" and to look at "transgovernmental" relations, that is, interactions among *subunits* of different governments which result in policy coordination and the building of transgovernmental coalitions to influence governmental decisions. International organizations—both IGOs and INGOs—may become partners in those transgovernmental coalitions; according to Keohane and Nye (1974: 52), "we would expect international secretariats or components of secretariats to form explicit or implicit coalitions with sub-units of governments as well as with nongovernmental organizations having similar interests."

Such transgovernmental, transnational coalitions do seem to have characterized the moves which led to Stockholm and Habitat, and persons associated with ISPAs appear to be among the coalition participants. Those networks seem to link together officials of governments, members of IGO secretariats, representatives of INGOs and national NGOs (as well as some individual "notables" whose participation is attributable more to their personal reputations than to their connection with a particular institutional or organizational base). The interactions among such persons in governments, in the secretariats, and in the private sector may be facilitated by their lateral movement from one role to another, by shared interests, class and educational background, common managerial and technical skills, and perhaps also by a shared commitment to finding rational solutions to problems posed by increasing international interdependence. The extent to which ISPAs are plugged into such networks may be a major factor affecting their capacity to exert influence in the international political system.

APPENDIX A

ISPAs WHICH SENT OBSERVERS TO UNCHE AND HABITAT, BY FIELD

Name of Organization	*UNCHE* *Number of Observers*	*UNCHE* *Responded to Survey*	*HABITAT* *Number of Observers*	*HABITAT* *Responded to Survey*
SOCIAL SCIENCE (11)				
*Commonwealth Human Ecology Council	3	x	19	
International Federation for Home Economics			2	
International Political Science Association			1	
*International Social Science Council	1	x	3	
International Sociological Association	1	x		
International Statistical Institute	1	x		
International Union for the Scientific Study of Population	1	x		
International Union of Psychological Sciences	2			
*Society for International Development	2	x	2	
*Union of International Associations	2	x	1	x
*World Society for Ekistics	2		23	

Name of Organization	Number of Observers	Responded to Survey	Number of Observers	Responded in Survey
LAW, ADMINISTRATION (12)				
*Eastern Regional Organization for Public Administration	1	x	1	x
International Association for Water Law	2	x		
International Association of Democratic Lawyers	2			
International Association of Legal Science	2			
International Association of Penal Law	3	x		
International Bar Association	3			
International Commission of Jurists	2	x		
International Federation of Women in Legal Careers			1	x
International Institute of Public Finance	1	x		
International Law Association	2	x		
*International Union of Local Authorities	2	x	5	
World Peace Through Law Center	1	x		
PROFESSIONS, EMPLOYERS (13)				
Inter-American Federation of Public Relations Associations	1			
*International Council of Nurses	2	x	1	
*International Council of Societies of Industrial Design	2	x	2	
*International Federation of Business and Professional Women	2		1	x
International Federation of Surveyors			1	
International Organization of Employers	2			
*International Police Association	1	x	1	
International Public Relations Association	2	x		
*International Union of Architects	2		10	x
Open Door International	1			
World Federation of Scientific Workers	1			
World Medical Association	1	x		
*Zonta International	1		1	
ECONOMICS, FINANCE (2)				
European League for Economic Cooperation	2			
*International Union of Building Societies and Savings Associations	2		2	x
TECHNOLOGY (15)				
European Federation of National Associations of Engineers	1	x		
Inter-American Association of Sanitary Engineering			1	x
International Association Against Noise	2			
International Association on Water Pollution Research	1	x		
International Commission on Irrigation and Drainage	1	x		
International Council for Building Research, Studies and Documentation			1	
International Electro-technical Commission	2			

Name of Organization	Number of Observers	Responded to Survey	Number of Observers	Responded in Survey
*International Federation for Housing and Planning	3	x	4	
*International Organization for Standardization	2	x	2	x
International Union of Air Pollution Prevention Associations	2			
International Union of Producers and Distributors of Electrical Energy	3			
Pan-American Federation of Engineering Societies	1	x		
Scandinavian Council for Applied Research	2	x		
World Energy Conference	6			
*World Federation of Engineering Organizations	1	x	3	
SCIENCE (20)				
Association of Official Analytical Chemists	1	x		
Caribbean Conservation Association			3	
European Physical Society	1			
*Friends of the Earth	6	x	14	x
Institute of Ecology			3	
Institute on Man and Science			1	
International Association for the Physical Sciences of the Ocean	1	x		
International Association of Microbiological Societies	2	x		
International Astronautical Federation	1			
*International Council of Scientific Unions	12	x	3	
International Geographic Union	2			
International Society of Radiographers and Radiological Technicians	2	x		
International Society of Soil Science	1	x		
*International Union for Conservation of Nature and Natural Resources	6	x	3	
International Union of Geodesy and Geophysics	2			
Pacific Science Association	1			
*Sierra Club	10	x	9	x
Society for General Systems Research			1	
World Wildlife Fund	5	x		
HEALTH, MEDICINE (13)				
International Association for Accident and Traffic Medicine	2	x		
International Association of Logopedics and Phoniatrics	1			
International Association of Medicine and Biology of Environment			2	
International Commission on Radiological Protection	1	x		
International Diabetes Federation	1			
International Federation of Gynaecology and Obstetrics	2	x		
International Medical Association for the Study of Living Conditions	2			

Name of Organization	*Number of Observers*	*Responded to Survey*	*Number of Observers*	*Responded in Survey*
International Society for Prosthetics and Orthotics			1	
International Union of Nutritional Sciences	2	x		
*Medical Women's International Association	2		1	
*World Federation for Mental Health	1		3	
*World Federation of Public Health Associations	1		1	
World Psychiatric Association	1	x		

*indicates organizations represented at both UNCHE and Habitat.

NOTES

1. The results of the Stockholm survey are reported in Feraru (1974).

2. The ISPAs represented at the Preparatory Committee (the number of sessions attended are indicated in parentheses) were: International Union for the Conservation of Nature (4), International Council of Scientific Unions (3), International Federation for Housing and Planning (2), World Society for Ekistics (2), Eastern Regional Organization for Public Administration (2), International Union of Local Authorities (2), International Law Association (1), World Peace Through Law Center (1).

3. UN Doc. A/CONF. 48/PC, 9, p. 86; A/CONF. 48/13, pp. 49-50.

4. Reports submitted to the Conference Secretariat by INGOs, national organizations, universities, and individuals are listed in UN Doc. A/CONF. 48/13, pp. 38-48.

REFERENCES

EVAN, W. M. (1974) "Multinational corporations and international professional associations." Human Relations 27, 6: 605.

FERARU, A. T. (1974) "Transnational political interests and the global environment." Int. Organization 28, 1 (Winter): 31-60.

KEOHANE, R. O. (1975) "International organization and the crisis of interdependence." Int. Organization 29, 2 (Spring): 357-365.

--- and J. S. NYE, Jr. (1974) "Transgovernmental relations and international organizations." World Politics 27, 2 (October): 39-62.

National Academy of Sciences (1972) Institutional Arrangements for International Environmental Cooperation. Committee for International Environmental Program Report to the U.S. Department of State. Washington, DC: Author.

RUGGIE, J. G. (1972) "The structure of international organization: contingency, complexity, and postmodern form." Peace Research Society Papers 18: 73-61.

Union of International Associations (1974) Yearbook of International Organizations, 15th editon. Brussels: Author.
United Nations Centre for Economic and Social Information (1972) Environment Stockholm. Geneva: Author.
WARD, B. and R. DUBOS (1972) Only One Earth. New York: W. W. Norton.

9

THE POLITICAL SIGNIFICANCE OF PUGWASH

Michael J. Pentz and Gillian Slovo

The first "Pugwash" Conference of Scientists[1] in July 1957 has a complex historical background which has been recorded only partially (Rotblat, 1962, 1972; Burhop, 1961; Wooster, 1969; Netter, 1976). The "prehistory" of Pugwash can be traced quite clearly back to Hiroshima and the *crise de conscience* that it produced among scientists in general and atomic scientists in particular, and perhaps further to prewar thinking about the social function of science and social responsibilities of scientists. Certainly, the idea of a conference of scientists "to appraise the perils that have arisen as a result of the development of weapons of mass destruction" did not arise suddenly but as a result of a confluence of many currents of opinion and concern among scientists about the enormous power and danger of the weapons they helped produce. The American fight for civilian control of nuclear power carried out by organizations such as the Bulletin of the Atomic Scientists (BAS) and the Federation of American Scientists (FAS) was an expression of this concern (Smith, 1965), and in Britain, the Atomic Scientists Association (ASA) was similarly involved. At the international level, the World Federation of Scientific Workers

AUTHORS' NOTE: The study is supported in part by the Open University and in part by a Research Award from the Leverhulme Foundation. We wish to record here our appreciation of the help and encouragement in our Pugwash study which we have received from Professor Bernard Feld and from Professor Josef Rotblat.

(WFSW), one of the few world federations with affiliated organizations from both East and West that was able to survive intact the pressures of the Cold War, had been concerned with these problems since its foundation in 1946.

It would take a few years before these independent moves would come together and yield first the Russell-Einstein Manifesto (Rotblat, 1972: Appendix 1) and then the first Pugwash Conference. The major obstacle was the apparently unbridgeable gulf between scientists from East and West, especially between Soviet and American scientists.

The Bikini hydrogen bomb test (March 1954) and the injuries produced by its fallout accentuated the concern about the implications of thermonuclear weapons. While various organizations and individuals discussed the possibility of more international discussions, the United Nations Organization had accepted a proposal to organize an intergovernmental conference on atomic energy.

Despite appeals to the contrary, by January 1955 it was clear that the conference would be confined to the peaceful uses of atomic energy and would furthermore be limited to official delegations from UN member states (thereby excluding China). While this UN move was accepted by some scientists in the hope that it would achieve something, the individuals and organizations most closely concerned up to this point went ahead with their efforts to convene an international meeting at which nuclear weapons could be discussed. The main streams of the many-channelled confluence were thus close to merging. The decisive factor which brought them together was the response to a Christmas broadcast made by Bertrand Russell in which he evaluated the dangers of nuclear weapons. The outcome of extensive correspondence following this broadcast between Russell, Albert Einstein, and others, and of meetings between Russell, J. Rotblat (ASA) and E.H.S. Burhop (WFSW) was the Russell-Einstein Manifesto[2] which, issued at a crowded press conference in London on July 9, 1955, began by calling for a conference of scientists to "issue an objective statement on this matter, addressed to a very wide public, over the signature of scientists of great eminence and of such a broad range of views that it will be possible to raise a cry of alarm without any section of public opinion being able to doubt the sincerity of the warning."

With this declaration, the basis for a conference was established and Russell, Rotblat, Burhop, and C. F. Powell formed themselves into a committee to organize it. After various financial setbacks, the committee accepted the offer of Cyrus Eaton to hold the first conference at his hometown in Pugwash, Nova Scotia, and the first conference was convened on July 7, 1957.

As we attempt now to estimate the significance of Pugwash over twenty years later, it is essential to appreciate the political conditions in which the protagonists of the first conference strove to bring together scientists from East and West. These were times of acute international tension, of cold war politics and unlimited development and testing of nuclear weapons. The political atmosphere today is very different, even though the perils of nuclear war are certainly no less than they were in 1957. The change from the politics of cold war to the politics of detente is, however, recent, partial, and not yet reflected in any significant measures of disarmament or of military detente. Nevertheless, the change is of significance to the history of Pugwash. Furthermore, the changed political circumstances have also led to a questioning of the future role of Pugwash, which is clearly reflected in the comments we have received from Pugwash participants in the course of our present study.

CONTEXT OF THE CURRENT STUDY

The Office of the Pugwash Continuing Committee issued a statement in 1974 which contains the following:

> It is in the nature of the Conferences that it is not easy to demonstrate their achievements or the success of the informal exchanges to which they have led. It has, however, been reliably stated that they made a very significant contribution to decisions by governments on such a major issue as the Partial Test-Ban Treaty. They have also contributed new ideas towards the control and limitation of armaments such as 'nuclear-free zones' and the 'black-box' methods of monitoring underground test-explosions. They may have been useful in preparing the ground for the recent 'SALT' negotiations.
>
> The Pugwash Movement played a special role in promoting international scientific exchanges on grave public issues, and its initiative has been followed by the creation of several expert institutions devoted to the study of resolution of conflicts and arms control. Nevertheless, the uniqueness of the Pugwash approach, the international confidence in its role, independence and integrity, and the mutual trust built up over the world, all demand that it should continue and intensify its efforts.

We have set out to seek answers to the questions: How influential has Pugwash in fact been? Has its influence grown or declined since the early years? What precisely have been the mechanisms by which it has influenced policy, if indeed it has at all?

METHODS OF INVESTIGATION

In our study of Pugwash we are using the following methods of investigation.

(1) A study of the archives and documents made available to us by the institution itself.
(2) A study of the archives of leading participants.
(3) A questionnaire addressed to the 800 people who have participated in Pugwash Conferences and/or Symposia.
(4) Personal interviews with individual scientists who are or have been active in Pugwash.
(5) Following the leads obtained from the above, particularly items 3 and 4, correspondence and/or interviews with politicians said to have been influenced in some way by the work of Pugwash.
(6) A search for corroboratory evidence in the historical records—e.g., of the U.S. Senate and Congress, and in UN documents.

At the present, intermediate stage of our enquiry, we have done a fair amount of work on the first four lines of approach, but nothing at all on the last two. We are thus not yet in a position to answer the questions posed above, although some hypotheses can be formulated at this stage.

SURVEY OF PARTICIPANTS

It seemed to us that the first step in assessing the influence of Pugwash was to ask its members what they thought about it. We accordingly drafted a questionnaire aimed at finding participants' views of:

(a) the effects of Pugwash on disarmament negotiations, attitudes of governments, scientists and the public, and on international scientific cooperation;

(b) the mechanisms or processes whereby Pugwash seeks to influence policy;

(c) any changes that have occurred (or should occur) in the above since the first conferences; and

(d) whether the influence and significance of Pugwash is perceived differently by participants from different countries.

We also asked for a general assessment of the utility of Pugwash, for suggestions about political contacts we might make for further information and for any additional comments.

In what follows we have quoted quite extensively from the questionnaire replies. When participants have given their permission, we have attributed the statement to them. In cases where the participant gave permission to quote them anonymously we have done so, giving their country only.

Table 1 shows how the 343 replies were distributed among different countries. Evidently the number of responses from the "developed capitalist countries" are high enough to permit some sort of statistical analysis, but the numbers from the "developed socialist countries" and from "Third World" countries are far too small for this. On the other hand, we received many thoughtful and interesting comments from participants in all parts of the world, including those from which the quantitative response was poor. The responses (and the comments) showed that some of the questions could have been formulated better than they were, and these defects naturally made interpretation of the results more difficult than it should have been. The poor response from the socialist countries was, of course, regrettable. We believe, however, that it reflects not an attitude to Pugwash but an attitude to questionnaires.[3]

FINDINGS

PERCEPTIONS OF THE EFFECTS OF PUGWASH

The first question on the survey concerning Pugwash was aimed at assessing the participants' view of its effects. It read as follows: "Do you think the Pugwash Conferences have had a significant effect on:

(a) disarmament negotiations

(b) your government's attitude to disarmament

(c) international agreement on arms control measures

(d) international scientific cooperation

(e) attitudes of scientists in your country

(f) attitudes of the general public in your country?"

The options given were yes, possibly, and no. The results are summarized in Table 2.

We think that the following tentative conclusions may be drawn from the responses to this question:

(1) More than two-thirds of all the respondents have a positive view of the overall effects of Pugwash (38% replying *Yes* and 31% replying *Possibly,* on the average, to the six questions, and only 24% replying *No*).

(2) There is most confidence about the effect of Pugwash on item d, international scientific cooperation and only slightly less on item a, disarmament negotiations; item c, international agreement on arms control, and item e, attitudes of scientists. But respondents are significantly less confident about item b, effect on their own governments' attitudes to disarmament, and very much less about item f, attitudes of the general public. However, the low "yes"

TABLE 1 Response to Questionnaire

Country	*Number sent*	*Number returned*	*% return*
USA	179	100	56
UK	77	40	52
Other DCC[a]	296	152	51
DCCX[b]	373	192	51
All DCC[a]	552	292	53
All DSC[c]	208	30	14*
Others[d]	122	21	17
ALL	882	343**	39

(a) DCC: Developed Capitalist Countries: Includes Western Europe, Canada, Australia, New Zealand, and Japan.
(b) DCCX: Developed Capitalist Countries except USA.
(c) DSC: Developed Socialist Countries: Includes USSR, Poland, Czechoslovakia, Hungary, Bulgaria, Rumania, and the German Democratic Republic.
(d) All Third World countries.

* Questionnaires were sent to 72 Soviet participants. There were only 7 individual replies. But a collective reply from the Soviet Pugwash Committee was sent by its Chairman, Academician M. A. Markov. The return from the other Socialist countries (23 out of 136) was only slightly less disappointing.

** 43 of the returns were blank. Reasons for this included ill health, lack of knowledge about Pugwash and, in a few cases, dislike of the questionnaire.

TABLE 2 Perceived Effects of Pugwash

	% responding "Yes" (Y) and "Possibly" (P)											
	US		*UK*		*DCCX*		*DCC*		*DSC**		*ALL*	
	Y	*P*	*Y*	*P*	*Y*	*P*	*Y*	*P*	*Y*	*P*	*Y*	*P*
(a) disarmament negotiations	54	37	46	46	40	49	44	45	51.9	33	46	43
(b) government's attitude to disarmament	40	36	30	51	11	40	20	39	33	42	21	39
(c) international agreement on arms control	54	37	58	38	46	47	44	44	42.3	42	44	44
(d) international scientific cooperation	66	27	51	31	55	35	58	32	73	27	58	33
(e) attitudes of scientists	63	25	41	48	39	42	47	36	54	37	47	36
(f) attitudes of public	9	34	3	24	6	28	7	30	40	24	10	28
Average (a) - (f)	48	27	38	40	32	40	37	37	49	34	38	31

*In view of the small numbers (about 25), these percentages are less significant than the others.

response for item b is due mainly to the less positive attitude of participants from countries other than the US, the UK and, perhaps, the socialist countries. This, together with the higher ratings for items a and c, suggests that participants from the non-"Great Power" countries tend in the main to see the need and possibility for a change in government attitudes to disarmament (and hence the possible effect of Pugwash) to apply to the major military powers. The low "yes" response for item f reflects a general view (supported by many of the written comments) that Pugwash has done very little to affect public opinion. Though the figures from the socialist countries are unreliable because the numbers are small, the relatively high rating for effects on public attitudes is consistent with what is known about the much greater publicity given in these countries to the concerns of Pugwash.

(3) It is noteworthy that U.S. participants are more positive in all aspects than participants as a whole, and more positive than any other group except, perhaps, in three instances. These are in respect of item c, where the United Kingdom response is comparable to that of the United States, item d, where the DSC response may be comparable, and item f, where the DSC response is perhaps significantly more positive. Whether this merely reflects the numerically greater involvement of American scientists in Pugwash (20 percent of all participants have been from the United States alone) or whether it reflects a greater involvement of American scientists in politics than in other countries, or at any rate other capitalist countries, remains an open question.

A closer scrutiny of the responses to part b (effect of Pugwash on "your government's attitude to disarmament") showed some significant differences between countries. While 41 percent of the U.S. participants (29 out of 71) said *Yes* to this option, only 1 of the French participants (out of 18) gave this response.

One of the problems with this question was that it referred to the Pugwash Conferences throughout their history. However, later comments by many of the participants point to a change in both the goals and effects of Pugwash over the years. In the beginning the membership was predominantly American, Russian, and British at a time when the informal contributions of Pugwash to disarmament/arms control and related matters were more necessary than now. Furthermore, the first conferences took place at a time when top scientists in the United States and the USSR were being involved in government decisions (e.g., Wiesner in the United States and Feodorov in the USSR). We accordingly sorted the responses into two groups: (A) those who have been involved with Pugwash before 1962; (B) those who have been involved only since 1962. The choice of 1962 as the dividing line was prompted partly by some of the comments and partly by the events of the 1961-1963 period.[4] The results are shown in Table 3.

In practically all respects, the pre-1962 respondents are more positive than the post-1962 respondents, but there is not much difference in this respect between the U.S. respondents and the ensemble. As previously noted, U.S. respondents are more positive than the ensemble irrespective of whether they are in the pre- or post-1962 groups. Taking the respondents as a whole, the main differences between pre-1962 and post-1962 indices relate to item a, disarmament negotiations ($\Delta = 18\%$); item b, government attitudes to disarmanent ($\Delta = 40\%$); and item f, public attitudes ($\Delta = 33\%$). For the U.S. respondents alone, the above three differences are about the same as for the ensemble, but in addition there are differences in relation to item c, international agreement on arms control (Δ 17%), and item d, international scientific collaboration. Probably the most significant difference between the pre-1962 and post-1962 groups relates to their estimate of the effect of Pugwash on their own government's attitude, which is distinctly more positive pre-1962 than post-1962.

A factor which might have affected the U.S. response pattern is the decline in influence of scientific advisers (and hence of the U.S. Pugwash group) in government circles after the death of President Kennedy and particularly during the presidency of Richard Nixon. Several comments refer to this trend. However, the contrast between the fairly positive response to part b by the Americans and the very negative response of the French probably reflects the French government's intransigence on disarmament in general and nuclear weapons in particular. As two participants commented:

> "My government is an heir of de Gaulle and thinks more of the 'greatness of France' than of disarmament." [J. Moch, France]

> "My government has had a stubborn negative attitude towards disarmament for the past 16 years." [F. Perrin, France]

The comments are typical of a feeling expressed in many of these comments: that the conferences *do* contribute something in situations where a need is perceived (e.g., at a time where there was no other channel for the Americans and Russians to talk about disarmament and related matters), but that the odds are against those who try to work in an atmosphere where both the government and the mass of scientists are uninterested.

PERCEPTIONS IN CHANGE IN THE EFFECTS OF PUGWASH

For the first Pugwash Conference, the fact that scientists from Russia and America could discuss sensitive political subjects was an achievement

TABLE 3 Differences in Perceived Effects of Pugwash Between Pre-1962 and Post-1962 Participants

	*ALL RESPONDENTS** *Positive Index***			*U.S. RESPONDENTS** *Positive Index **		
EFFECT ON	*Pre-1962* %	*Post-1962* %	Δ*** %	*Pre-1962* %	*Post-1962* %	Δ %
(a) disarmament negotiations	52	44	18	54	47	15
(b) government's attitudes to disarmament	35	25	40	46	35	31
(c) international agreement on arms control	50	46	9	55	47	17
(d) international scientific cooperation	52	53	-2	66	57	16
(e) attitudes of scientists	44	44	0	52	52	0
(f) attitudes of public	16	12	33	18	14	29
Average (a) - (f)	41	37	11	47	42	12

* The numbers were: All respondents: Pre-1962 = 50, Post-1962 = 134; U.S. respondents: Pre-1962 = 54, Post-1962 = 47.

** Calculated as (3 × %'Yes' + %'Possibly')/4.

*** Δ = 100 (%Pre-1962 – %Post-1962)/%Post-1962. This index, rather than a simple percentage difference, was used in order to adjust for the difference number of participants in each group.

in itself. In the following question, we attempted to evaluate the participants' perceptions of any change that might have occurred since then: "Do you think that since the first conference there has been a change in the effect of the Pugwash Conferences on:

(a) scientists in your country

(b) your government's attitude to disarmament

(c) disarmament negotiations?

If so, please specify." The options given were Yes or No. The results are summarized in Table 4.

As can be seen from Table 4, a significant proportion of the participants do see a change since the first conference. Once again there are differences between the United States and the United Kingdom and the other countries. Thus, for instance, a higher proportion of participants from the United States and United Kingdom than from the other capitalist countries perceive a change in the effect on scientists. (Due to the small numbers the DSC percentages are not reliable.) There is an even more pronounced difference in the perceived effects on government attitudes, whereas there are no significant differences about effects on disarmament negotiations.

Not all those who saw a change specified its direction, but for those who did we have grouped their comments as follows:

Change in the Effect on Scientists

Of 115 specific comments on the effects of Pugwash on scientists, 62 felt that there had been an *increased* effect, 33 that there had been a decreased effect and 20 were ambiguous.

Some examples of comments from those who saw an *increased* effect were:

> "Pugwash has become more popular among the scientists of my country." [W. Frank, Austria]

> "Especially in the degree of knowledge about disarmament matters and in understanding the necessity to cooperate in this field." [Poland]

> "Wider influence among *social* scientists." [United Kingdom]

> "Greater awareness of the political difficulties of arms control." [W. A. Higinbotham, United States]

> "U.S. scientists have been the principal source of initiatives in this country for disarmament. I think Pugwash helped enlist them."

TABLE 4 Perceived Change in Effect of Pugwash Since the First Conference

There has been a change in the effect of Pugwash on:	*% OF POSITIVE RESPONSES*					
	US	*UK*	*DCCX*	*DCC*	*DSC*	*ALL*
(a) scientists in your country	84	80	58	66	77	66
(b) your government's attitude to disarmament	73	71	34	46	33	43
(c) disarmament negotiations	68	75	65	65	58	64

Another view expressed by some of the respondents (and inevitably contradicted by others) is that the image of Pugwash has changed from that of a "radical" (and hence dangerous) organization to one which would be more acceptable to the majority of scientists:

> "Pugwash has become better known and is no longer seen as populated by dangerous radicals." [Ph. B. Smith, Netherlands]

> "I think that the Pugwash Conferences have come to be more regarded as useful and serious—not just as a 'radical' or 'fringe' movement." [E. Staley, United States]

Examples of the comments from the significant minority who saw a *decrease* in the effect of Pugwash on scientists in that country can be sorted into three groups:

(i) Those who stress that the initial challenge of the Pugwash movement has diminished either because the threat of war has receded or because Pugwash itself helped open new channels which now exist independently of it:

> "On the one hand, there has been dissemination of information on the conferences, while, on the other, there has been less interest shown by the postwar generation of scientists who do not have a vivid memory of war." [Y. Sakamoto, Japan]

> "A declining impact due to apparent detente and less concern over nuclear war and a multiplication of organizations of a different nature concerned with the responsibilities of scientists." [United Kingdom]

> "Early impact [was] significant as it reinforced ideals of political comment and opened international channels. Norms are now established independent of Pugwash." [United States]

(ii) Some responses were specific to certain countries and tend to reflect the general attitude of scientists in those countries which then determines their views of Pugwash. For example, an Austrian participant writes:

> "To my regret, scientists in Austria are still not enough interested in world affairs and look at the activities of Pugwash with a certain amount of scepticism."

(iii) In contrast to those who see the more respectable image of Pugwash as giving it scope for greater involvement of scientists, some participants see this as a cause for a waning of interest:

> "At first regarded as 'leftist,' then respectable enough for the bolder Establishment scientists, now probably too respectable." [N. Calder, United Kingdom]

Change in Effect on Government's Attitude to Disarmament

The comments on this point were almost evenly divided between those who saw an increased effect, a decreased effect, or no net change. Perhaps predictably, several comments referred to the relative powerlessness of scientists in determining their government's attitude to disarmament. The emphasis is not on the worth or possibilities of the Pugwash Conferences, but rather on the external and internal pressures on government which determine its disarmament policy. Thus there were a series of statements of the type:

> "Government follows policy requirements seen as usual." [B. Kind, Federal Republic of Germany]

> " . . . always following his big Western brothers." [J. Galtung, Norway]

> "The attitude of my government—the government of a neutral country for which heavy weapons (missiles and so forth) are forbidden by the states treaty of 1955—towards the question of disarmament is a passive one." [W. Frank, Austria]

To many Americans, the date of the decline of influence of the Pugwash Conferences is closely tied to the waning influence of scientists within government which occurred in the last 15 years (and especially during the Nixon Administration):

> "At the outset I reckon that Pugwash scientists were a strong pressure group towards disarmament: primarily because some of the U.S. participants were politically powerful. This is certainly not the case now." [M. Eden, United States]

For those who did see a definite change in the effect of Pugwash, the institutionalization of channels which Pugwash helped establish is given as a cause for the decline of the status of Pugwash:

> "The first few conferences had an effect and helped to get official negotiations going. Now there is a regular governmental organization, I doubt if Pugwash has much effect." [E. Bullard, United States]

> "I think the first conferences were the most effective. Later, gradually more and more factors began to work, and among them the relative importance of the Pugwash conferences decreased." [Hungary]

The other side of the negative effects on Pugwash of the acceptance of its ideas is that the conferences have become more respectable and are therefore increasingly listened to by governments:

> "I suspect the subject is now viewed as politically less hazardous than it was in 1959." [D. F. Cavers, United States]

Some respondents pointed to the greater trend to take scientists seriously:

> "Greater inclination to take into account certain proposals and ideas sponsored by scientists." [Sweden]

And, finally, one of the interesting things about the comments on this section is the way that influence exerted on governments is often seen in terms of participation by the *individual* who brings ideas from Pugwash into government but who acts separately from the organization:

> " 'Alumni' of Pugwash Conferences became advisors to government." [United States]
>
> "Influenced through discussions by Senators and Congressmen with Pugwashites." [United States]
>
> "By some influence of individual scientists in our government." [D. von Ehrenstein, Federal Republic of Germany]

This sort of viewpoint is worth noting since it is consistent with the Pugwash image of the independent scientist establishing international contacts, discussing matters with counterparts in other countries, and then using personal influence in the scientist's own country to spread Pugwash ideas. It is a view which some participants deny: quite a few of them comment later in the questionnaire that the Russian participants were always seen as official representatives of their government and that, once the Americans became aware of this, it became true for them too (although many who hold this view tend to see it applying in the early years—up to about 1962—of Pugwash's involvement).

Change in the Effect of Pugwash on Disarmament Negotiations

About one-half the comments indicated an *increased* effect since the first conference and about one-fifth indicated the opposite view. For the remainder, about one-third, the effect has waxed and waned somewhat. We suspect that for this part of the question even more than for the first two parts, responses and comments were rather strongly influenced by the respondents' judgment of the disarmament negotiations themselves rather than the increased or decreased influence of Pugwash upon them.

PERCEPTIONS OF THE IMPORTANCE AND INFLUENCE OF PUGWASH

After the first Conference, the newly elected continuing committee saw the aims of Pugwash as threefold: to influence government, to form a channel of communication among scientists, and to educate public opinion. We tried, by asking participants to rank various options, to assess their views of the success of these aims and to get their impressions of

Pugwash's importance and sphere of influence. The rather complex three-part question on this yielded an enormous amount of information. Rather than present the extensive data and complicated statistical tests that go with it, we give below a short summary of the main points.

The consensus view in this question is that Pugwash's main influence has been on "understanding between scientists of different countries," that their importance lies in the "free exchange of views on world affairs between scientists of different countries" and that this has been possible because of (1) the expertise, and (2) scientific rigor of the participants (the latter being considered marginally less important). At the other end of the scale is the impression that the Pugwash Conferences have had least recognition from, and influence on, the public.

IMPACT ON GOVERNMENTS

The evaluation of the relative impact of the conferences on governments highlighted some interesting differences among the participants. Although the consensus was that the conferences owed very little to their *participants'* direct access to government, many respondents felt that the importance of Pugwash lay in its influence on *scientific advisers* to government—and especially the governments of the United States and USSR. For the American participants this was particularly true—they ranked the influence on scientific advisers above the exchange of views between scientists. In addition, the Americans regard access to the *scientific community* as marginally the most significant reason for Pugwash's achievement, while the "ability to discuss problems rigorously," which was rated very highly by other participants, was given the lowest priority. It is evident that the "influence links" appear to be more significant to Ameri-

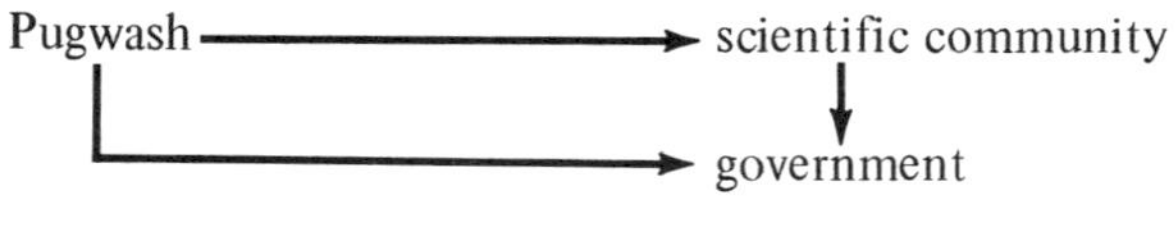

can scientists, who seem much less impressed than their colleagues by the ability of scientists to discuss problems rigorously and with expertise.

PERCEPTIONS OF POTENTIAL EFFECTS OF PUGWASH

Participants were asked whether the Pugwash Conferences could still make a significant contribution to: (a) disarmament negotiations, and (b) problems of developing countries. There were no significant differences in the responses to this question from participants of different countries. It is clear that a large majority of the respondents believe that Pugwash still has

TABLE 5 Perceptions of Potential Contributions of Pugwash

	ALL COUNTRIES		
	Yes	*No*	*Total*
(a) disarmament negotiations	182 (76%)	56 (23%)	238 (100%)
(b) problems of developing countries	179 (75%)	58 (24%)	237 (100%)

a significant contribution to make in both areas (see Table 5). Other areas in which Pugwash could make a contribution were suggested by 125 respondents. Their comments were necessarily brief, and in the main fall into the following three categories:

(1) Problems of ecology, energy, pollution, nutrition, raw materials, environment, developing nations, and population; the convergence of global problems in general.

(2) Uses of science for the benefit of mankind. Pugwash could make a contribution by:

   (a) discussing the technical applications of science and promoting international political measures for the regulation and control of the effects of technology;
   (b) spreading information to the general public; and/or
   (c) discussing and recognizing the responsibility of scientists: coming to a mutual understanding about this and informing and educating other scientists.

(3) Prevention of conflict by:

   (a) discussing and contributing to the solution of regional security problems;
   (b) developing international contacts among scientists;
   (c) promoting negotiations between countries in conflict; and/or
   (d) deepening the "detente" process, especially in Europe.

Some respondents said that they see no possibility of a Pugwash contribution in any field. Some do not see a "movement" of scientists but merely that individuals take part in Pugwash for their own benefit. Others state that scientists as such do not have any special claims to being able to solve any of the problems by virtue of their expertise.

ATTITUDES TOWARD THE BROADENING OF PUGWASH OBJECTIVES

The participants were asked for their impressions of the effect of the change in emphasis in the Pugwash Conferences away from disarmament and towards other problems, such as population, underdevelopment, and so forth. We received 204 comments on this question, ranging from single words (such as "good" or "negative") to long letters, reprints of articles, or texts of contributions made at Pugwash meetings. The comments were equally divided between approval (72) and disapproval (70) of such change in emphasis as there has been, with an almost equal number (62) of neutral, noncommittal, or null comments (e.g., advantages and disadvantages of the trend given about equal weight). We have divided the comments into five categories:

(1) Disapproval on the grounds that problems of underdevelopment, population, and so forth are already covered—often more completely and efficiently—by other bodies.

> "The change in emphasis has obviously led the Pugwash Conferences into broader fields, where already a great number of other scientists etc. have made their contributions. The voice of the Pugwash Conferences is therefore one voice amongst many, while in the opening phase, they were alone." [K. Evang, Norway]

The impression is one of the Pugwash Conferences' adopting issues which have become fashionable but without the original impetus which made them concentrate on disarmament. In addition, many participants feel that Pugwash has a special role to play in disarmament discussions while in other fields it can only act as other organizations:

> "The effect has been decreased influence due to diversification and thus weakening of effort. Also, while Pugwash has a unique standing in disarmament circles this is not so with respect to the other fields where a number of organizations existed before Pugwash entered them." [P. L. Olgaard, Denmark].

Those who answered in this vein do not necessarily consider the issues in themselves to be unimportant; rather they feel that Pugwash can only lose by spreading into fields which are already cause for comprehensive research and debate.

(2) Some scientists took this further. In addition to speaking about the organizations which work in fields other than disarmament, they point to the inability of Pugwash scientists to discuss these issues on the same level as they could with disarmament problems:

> "Largely to discredit the standing of the Pugwash Conferences. Population, underdevelopment, environment, energy, etc. have generally been handled poorly by Pugwash, producing some vague generalities, vows to abolish sin, empty rhetoric, and hot air than in the earlier days of the disarmament conferences, where topics were *specific* and discussed by experts." [United Kingdom]

Participants who commented in this vein see a "weakening of rigor" and a loss of the old Pugwash expertise. In connection with this, some people stress the fact that while discussions on disarmament implied widespread technical problems to be solved, the problems of underdevelopment are seen as far more general (political):

> "My first impression . . . is that the new emphasis on the problems of development has reduced the capacity of Pugwash to deal with problems in a technically sound way avoiding conventional political skirmishes. I am afraid that the Pugwash (movement) begins to suffer some of the problems of UN bodies and Agencies." [C. Schaerf, Italy]

This criticism seems to stem from a very definite view of the role of Pugwash. It is one where scientists supply their special knowledge on a subject where the facts can be identified and discussed: where there is a role for the expert to clarify the situation. Once Pugwash steps outside of problems such as these, it is suggested that scientists no longer have this special role but instead take their place alongside those other agencies which are struggling with the very complex problems of development.

(3) Some participants disapprove of the change of emphasis purely because it diverts attention from the problems of war. To those who replied in this way, it is not that development and population are subjects not suitable for discussion, but rather that the dangers of war will be ignored by governments and the public if Pugwash does not continually point to them:

> "One (negative) side effect may have been, that people believe the armaments race under control or even disarmament well under way. On the other hand the public attention is drawn to the not much less urgent problems of our time." [P. Blau, Austria]

(4) To some respondents, the change is not a good one because it carries with it less Russian interest and involvement. This view has to be tied to a recognition of the origins of Pugwash; of a time when Pugwash was virtually the only meeting place for Russian and American scientists to discuss the problems of nuclear arms in a dispassionate way:

> "In my view, Pugwash was important because it gave Russian and American scientists a respectable meetingplace in the cold war period. The later topics of discussion, while important, do not need respectability." [J.D.B. Miller, Australia]

and S. Bjornholm (Denmark) comments:

> "It carries the risk of diminished interest in Pugwash by USSR."

It is difficult to assess the basis for statements like those given above because of the low Soviet response (see Table 1). However, in a letter and in a number of papers sent with it by M. A. Markov on behalf of the Soviet Pugwash Committee, there was some indication that this feeling might be justified. Although the Soviet Pugwash group does not think that problems of development are unimportant, they clearly believe that the main emphasis of Pugwash should continue to be on the question of disarmament and detente, as the threat of nuclear war has by no means been removed.

(5) Some of the respondents, while recognizing the change in emphasis as a not unmixed blessing, tried to explain the reasons for it. To them, the change was inevitable—a reflection of changing priorities in the world which, in many ways, made the old ideas of Pugwash less relevant. To these people, the accent is on the necessity of the change imposed from outside rather than on an evaluation of whether it was a good/bad move for Pugwash:

> "Positive effect. The Pugwash movement may now have outlived its usefulness—or at least passed its zenith—in the field of disarmament. I think it is suitable to use the apparatus which has been established over the years in problems which may be more important than disarmament at present." [N. Peterson, Denmark]

For some, the change has not been a total success and yet the blame is not attached to Pugwash but rather to the changing place of scientists within government.

> "It is a natural reflection of the changing priorities of the contemporary world. Unfortunately, Pugwash has no adequate competence to deal with these problems." [Y. Sakamoto, Japan]

> "A positive effect in the sense that scientific arguments enter more sectors of world politics. Widening the scope of activities has not caused the decline in importance of Pugwash in disarmament. That would have occurred anyway." [J. Prawitz, Sweden]

Most of the above judgments have been negative about the change in emphasis. There were basically three broad reasons why some respondents consider the change to be a good one:

(a) that these new issues are vitally important for peace;

(b) that different scientists are drawn into the movement which now talks about problems recognized as important by the public; and/or

(c) that the change has opened the way for discussions on the uses/abuses of science and forced scientists to recognize more responsibility for their work.

View (a) (that other problems are important for world peace) is based on the recognition that disarmament cannot be considered in a vacuum or that the arms race is not the only threat to the world. Those people who answered in this category talk about the interconnection of the need for more equality with disarmament. The idea is that poverty itself leads to war:

> "Very welcome effect—because the root causes of war are poverty—the seeds of peace are in nutrition, education, and development." [R. P. Devadas, India]

> "The change was necessary because disarmament is not an isolated problem and cannot be solved by itself." [B. Bertotti, Italy]

To some participants the change in emphasis of view (b) has been a good one, since it has opened Pugwash up to the general public, to a broader circle of scientists, and to the governments of the Third World:

> "I suggest that change to problems of population and others of developing countries has increased . . . the influence of Pugwash [on] the public." [H. Kraut, Federal Republic of Germany]

> "It has brought in new people . . . and has broadened the appeal of the movement to the scientific community and the public. It has enlisted more social scientists, which I think is a good thing." [E. Staley, United States]

> "Broadening the range of people who can attend Pugwash meetings and the range of impact of Pugwash on different government departments and agencies and international agencies." [J. H. Chesshire, United Kingdom]

Finally, there are those expressing reason (c), who point to the even wider issues that the discussion on development has provoked. This view suggests that Pugwash participants have been forced to think about the problems politically and reassess their place in what was previously regarded as purely technical discussions:

> "Confronting more scientists with the necessity of taking political responsibility for their work and of viewing direct violence within the context of structural violence." [L. Brock, Federal Republic of Germany]

"A good effect—the broadening of the field of responsibilities and activities of scientists of different disciplines." [I. Kende, Hungary]

THE PUGWASH MOVEMENT

A UK participant commented:

"It is extraordinarily difficult to estimate the effects of a movement such as Pugwash. I believe, however, on general grounds, that its influence has been greatly to the good and that it is better to spend time continuing this work than in trying to measure the immeasurable."

It seems to us, however, that some attempt at least to *estimate* the political effectiveness of Pugwash and to identify the factors which have been mainly responsible for such effectiveness as it may have had, could be of some importance in "continuing this work." Several comments underline the *indirectness* of the political impact of Pugwash. Another example of an indirect effect of Pugwash is the establishment as an offshoot of Pugwash of the SADS group (the bilateral U.S.-Soviet study group on arms control problems, comprised of people from the Soviet Academy of Sciences and the American Academy of Arts and Sciences) which continues to meet and provide an informal East-West channel on this major problem. In a letter enclosed with the questionnaire, Talcott Parsons makes this comment on the influence of Pugwash:

Though this movement did not receive very wide publicity, it did have access to top people in the scientific community on both sides, and, through them, less direct but none the less significant access to the decision-making people in their respective governments. It seems to me that it has turned out to be a movement of very substantial significance in helping to stabilize a very unstable situation which followed World War II.

Writing in the Bulletin of the Atomic Scientists, Eugene Rabinowitch (1965: 10) sums up many of the comments we received:

The Pugwash discussions have contributed to the acceptance by the Soviet leadership of the need to consider first the early, practically possible steps towards disarmament, and to relax their original all-or-nothing attitude. They have contributed to the acceptance by both sides to the principle of minimum deterrence. . . . The Pugwash Conferences have helped both sides to understand that the other has legitimate security requirements, the insistence on which does not necessarily mean the absence of a genuine desire for disarmament. They have helped to explore compromise solutions in deadlocked controversies on disarmament controls.

More specifically, it is claimed that Pugwash Conferences contributed significantly, although indirectly, to international agreement on the prohibition of biological weapons, to the Non-Proliferation Treaty, the SALT I agreements, the establishment of nuclear-free zones, and to the renunciation of ABM strategies. The difficulty in assessing these claims is in obtaining independent corroboration and in identifying in any but the most general terms the processes or mechanisms through which the Pugwash Conferences had such influence.

Eugene Rabinowitch (1965:10) goes on to cite as "perhaps the best illustration" of the function of Pugwash Conferences in exploring compromise solutions to deadlocked controversies on disarmament controls the example of the proposal made jointly by four Soviet and American participants at the 1961 Conference in Cambridge, England, for the establishment of unmanned seismic detection stations ("black boxes"). The proposal "was almost immediately picked up in official negotiations"—it "seemed to offer a mutually satisfactory answer to the controversial problem of adequate control of underground tests." The "black box" idea had in fact been mentioned some months earlier in informal discussions with academician E. K. Feodorov in Geneva. As Feodorov was at that time leading the Soviet scientific delegation to the Geneva test-ban talks and was also on the Pugwash Continuing Committee, it is not surprising that the idea, which was undoubtedly being considered by a number of people at the time, should have been put forward at Cambridge by academician I. E. Tamm.

Rotblat (1972: 55) records in his history of the Pugwash Conferences:

> This so-called 'black-box' idea received a great deal of publicity, as it was mentioned in the public statement of the Conference. It was soon afterwards taken up at government level, and figures in an exchange of letters between Kennedy and Kruschev [sic] a few months later.

This instance illustrates the complex (and often hidden) strands in the network of interactions between individual scientists, official negotiators, scientists' organizations (including Pugwash), and political decision makers, which characterize this whole area. These interactions seem to create an atmosphere conducive to successful political negotiation rather than achieving any obvious breakthroughs on specific issues, and it is this which makes it so difficult to assess the achievements of Pugwash.

In the field of international scientific cooperation (which has very clear links with wider questions of political detente), cooperation in space exploration (including common utilization of satellites for communications and meteorological observations) is cited as one example of a

Pugwash proposal that has been realized. Another, quite explicit, outcome of a Pugwash proposal is the setting up of the International Centre for Insect Physiology and Ecology.

Perhaps a comment by Eugene Rabinowitch (1965: 14) is a fair reflection of the consensus expressed in the answers we received to the final question in this questionnaire:

> The Pugwash 'movement' is much weaker than some people outside may think—but it is the only international movement trying to change national opinion and political leadership everywhere toward better understanding of the challenges of the scientific revolution and of policies adequate to cope with these challenges.

TENTATIVE CONCLUSIONS

We said in the introduction that we set out to seek answers to the questions: How influential has Pugwash been? Has its influence grown or declined since the early years? What have been the mechanisms by which it has influenced policy, if indeed it has? It must be emphasized that at this intermediate stage of our enquiry we are not in a position to attempt any definitive answers to these questions. Clearly it would not be valid to draw conclusions about the influence of Pugwash simply by asking the participants for their views on this question, unless the consensus were to be overwhelmingly *negative.* A positive consensus, such as has emerged from the survey (and also from the documents and interviews), could be suspected of reflecting an understandable tendency to rationalization and self-justification. We need to be cautious even about drawing too many conclusions from the questionnaire above, without supporting evidence from other sources, about what the "Pugwashites" themselves think about Pugwash. How representative is our sample of 300 or so of the population of over 800 Pugwash participants (or even the larger proportion of *American* participants)?

The responses to questions about the effect of Pugwash showed that a large majority of respondents believe that Pugwash *has* had a significant effect on *attitudes of scientists,* on *international scientific cooperation* and on *disarmament negotiations.* There is somewhat less confidence in the United States and the United Kingdom (and perhaps the Socialist countries) about the effect on their own government's attitude to disarmament, and very much less in other countries. And with the possible (and statistically dubious) exception of the Socialist countries there is a clear consensus that Pugwash has had no significant effect on public opinion.

At the time of the Fourteenth Pugwash Conference, on "International Cooperation for Science and Disarmament" (Venice, 1965: 14) Eugene Rabinowitch (1965: 14) wrote:

> While generating ideas by discussion in small 'thinking' conferences and study groups is the most important function of Pugwash, and refraining from large meetings and public propaganda has proved a wise decision, there is no doubt that Pugwash Conferences also have a function of influencing public attitudes, to make them responsive to new ideas.

In the view of our respondents at least, this function has been only partially fulfilled. That there should have been so little impact upon public opinion is, for some of our respondents more than others, a cause for concern. The differences on this may reflect differences in conceptions of what the role of the "scientist" is, whether the scientist has a special role at all, whether the impetus for arms control and disarmament comes from the individual expert or groups of experts or from a mass-based campaign. As one participant says:

> "It is hard to respond to questions regarding Pugwash in general. Different issues are handled very differently. One common thing, though, is the lack of understanding of the role of the general public. Real change, of course, must come from the bottom. In some cases, like possibly disarmament and arms control, workable lines of attack could be worked out at conferences like Pugwash's but they will then afterwards have to be introduced into some kind of mass activity." [B. Erikson, Sweden]

If the view that in the last resort it is public opinion that influences political decisions is accepted, then the doubts remain over the intermediate stages to this "last resort" and whether scientists can influence any part of the process.

There is general agreement among the respondents (as there seems to be among all those who wrote about Pugwash) that the Pugwash Conferences played an important role in opening up a channel of communication between the United States and the USSR in the late 1950s and early 1960s. One strongly supported view is that this original role of Pugwash to enable East and West to have informal friendly discussions is no longer needed.

> "Pugwash at the time it was organized was unique and very useful. Since then there have been many other opportunities for exchange between East and West, and so the usefulness of Pugwash was diminished." [A. M. Weinberg, United States]

The idea of "frank and informal discussions" was obviously a necessary and useful one which yielded results by reducing antagonisms, developing mutual understanding, allowing for information exchange and, perhaps, by changing government attitudes. The view, stated explicitly by many respondents, that Pugwash was more effective in this field up to 1962-1963 than it has been since, is reflected in the differences (see Table 3) between respondents who were involved with Pugwash before 1962 and those involved only since that time. But the most pronounced difference was shown by U.S. participants with respect to effects on their own government's attitude to disarmament, and this could have more to do with political trends since the death of President Kennedy than with Pugwash as such.

It is generally agreed that the influence of Pugwash has been mainly on mutual understanding *between scientists* of different countries, recognition *by scientists* of their responsibilities and, to a lesser extent, on disarmament negotiations and arms control measures. *Public* recognition of the importance of science in world affairs is again rated lowest. There is a consensus view that it is mainly through the free exchange of views on world affairs *between scientists* of different countries and through the influence of the conferences on *scientific advisers* to governments and of conference participants *on the governments of the United States and the USSR* that Pugwash has had effect. A number of specific examples are quoted (such as the Test-Ban Treaty) but, of course, without supporting evidence (which was not requested).

In spite of the clear impression that Pugwash is believed to have had significant effects upon "scientists," we are left with the question *which* scientists? Is the effect limited essentially to a small "elite," members of the club, mostly older generation, "establishment" scientists? Some of our respondents think that this does happen:

> "Pugwash has adopted too elitist an approach. The British Pugwash Group is kept firmly in its place. It really plays no role. In Britain, much more effort should be made to democratize Pugwash activities and to make all British delegates more responsive to modern thinking about science and its role. Younger people should be encouraged very much more to participate and there should be much more contact with organizations and scientists." [E.H.S. Burhop, United Kingdom]

We suspect that there may be significant differences here between the United States (and perhaps the United Kingdom and the USSR) and most other countries, and that one of the factors that might have contributed to this difference (for the United States) has been the influence of the

Bulletin of the Atomic Scientists and of Eugene Rabinowitch as its editor. Major efforts were also made by the Soviet Pugwash Committee (e.g., to publicize the Vienna Declaration) to inform Soviet scientists about Pugwash Conferences. But the effect, even on scientists (let alone politicians or the public), seems to have been much weaker, on the whole, in Western Europe, and practically negligible elsewhere, with the possible exception of Japan, where there are in any case other reasons than the "solemn warnings" of Pugwash scientists for general concern about the danger of nuclear war.

The established "Pugwash method" of small, private conferences based on working groups and avoiding the perils of omnibus "declarations" from plenary sessions, has obviously been necessary and effective. Attempts to widen the influence of Pugwash through the activities of the national Pugwash Committees (including study groups and regional conferences) seem to have had a positive but limited effect. It is possibly in these directions that Pugwash should in future be concentrating its efforts rather more than it has done so far, especially with a view to the involvement of younger scientists and to extending the scientific base of the "movement" beyond the natural sciences.

The questions about *changes* in the effects of Pugwash and in the degree of emphasis of the conferences on disarmament matters were clearly interconnected in the minds of many of the respondents, and the question of "diverting" some of Pugwash's efforts from problems of disarmament to problems of development raised a lot of comment—and revealed much divided opinion. A majority of respondents, varying somewhat according to their country, believes that there *has* been a change in the effect on scientists in their country, mostly positive. Similarly, a majority believes that there has been a change in the effect on their government's attitude to disarmament—though there is much greater variation between countries here, and there appears to be about equal partition between those who perceive an increase and those who perceive a decrease in effect. As for the effect on disarmament negotiations, a majority perceives a change since the first conference but there is again deep division on whether the change has been positive.

An inherent problem in assessing the political effects of Pugwash arises from the multiplicity of channels through which these effects are transmitted. Most "Pugwashites" are also members (often leading members) of other professional associations both national and international. These include bodies such as the Academies of Sciences, the Physical Society (and analogous professional associations), as well as other organizations and movements, such as Societies for Social Responsibility in Science and many others. Other Pugwashites are active in trade union organizations of scientific workers with very strong links to the rest of the trade union movement. (This applies to certain European countries and to the socialist countries but not to the United States.) At the international level there are important personal links with both intergovernmental and nongovernmental institutions. Leading members of Pugwash occupy important positions in such organizations as UNESCO, WHO, FAO, and others in the UN family, or in such nongovernmental organizations as ICSU, IUPAP, and analogous bodies. The ramifications of Pugwash extend even into such highly specialized international organizations as CERN, which can count among leading Pugwash members two of its former Director-Generals (Weisskopf and Gregory), several members of its Council and an appreciable number of its senior scientific staff. The links with the WFSW, through an unbroken succession of presidents and members of its Executive Council, have already been mentioned. Indeed, one way of estimating the interconnections of Pugwash with the rest of the institutions depicted in Evan's (1974: 592) "non-linear model of integration in the international system" might be to identify the organizational and personal links of individual Pugwashites with the rest of the network.

It may even be possible to devise a semiquantitative estimation, at least on a comparative basis. The informality of Pugwash, inherent in its whole philosophy and method, makes it much more difficult to trace attempts at influencing policy than is presumably the case with more formally organized institutions. The elusiveness of Pugwash to attempts to pin down either the degree or the mode of its influence on policy is its special characteristic and, ironically, may also be the key to its success. One of our commentators expressed the view that "Pugwash is a test case of the new view (on responsibility, accountability, ethical impact of scientific work) on the part of scientists and the public."

NOTES

1. The Pugwash Conferences on Science and World Affairs and the somewhat rudimentary organization responsible for them are known variously as "Pugwash COSWA," "the Pugwash Movement," or just "Pugwash." We shall use the last name, as it is the simplest.

2. The signatories were: Max Born, P. W. Bridgman, Albert Einstein, Leopold Infeld, Frederic Joliot-Curie, H. J. Muller, Linus Pauling, C. F. Powell, J. Rotblat, Bertrand Russell, and H. Yukawa.

3. The questionnaire was sent to the USSR and some other countries both in English and in Russian translation. We do not think the problem was one of language.

4. December 13, 1961: UN approved U.S./USSR principles for arms negotiations and creation of the Disarmament Committee. March 14, 1962: 18-Nation Disarmament Committee set up (later became CCD). March 15, 1962: USSR presents draft

treaty on general and complete disarmament (GCD). March 18, 1962: United States presents draft treaty on GCD. July 20, 1963: Hot line set up. August 5, 1963: Partial Test-Ban Treaty.

REFERENCES

BURHOP, E.H.S. (1961) "The origins of the Pugwash Movement." Scientific World 5, 3: 11-16.

EVAN, W. M. (1974) "Multinational corporations and international professional associations." Human Relations 27: 587-625.

NETTER, F. (1976) "Naissance et developpement du mouvement Pugwash." Le Monde Diplomatique 23, 270 (September): 18.

RABINOWITCH, E. (1965) Bulletin of Atomic Scientists 21, 4: 9-15.

ROTBLAT, J. (1972) Scientists in the Quest for Peace–A History of the Pugwash Conference. Cambridge, MA: MIT Press.

---- (1962) Science and World Affairs–History of the Pugwash Conferences. Pall Mall, London: Dawsons.

SMITH, A. K. (1965) A Peril and a Hope. Chicago: Univ. of Chicago Press.

WOOSTER, W. [ed.] (1969) A Short History of the WFSW. (unpublished)

10

THE CLUB OF ROME AND ITS POLICY IMPACT

Alexander King

The Club of Rome is a rather unusual body in its constitution, operation, and objectives. It is quite unlike the average nongovernmental organization and hence its evolution and approaches can only be understood if something is known of its origins and early development. Hence in this chapter I shall give a general account of how the Club arose, and how its motivations developed.[1]

The Club consists of a maximum of 100 members. This number has been selected as the maximum possible that can be serviced without a formal secretariat, because one of its main features since the beginning has been the concept of maintaining it as a club rather than an organization, with as light a structure as possible. It has, in fact, no formal secretariat and no budget. Its only committee is a small executive group of eight people who meet rather frequently and are the nucleus of its operations.

The 100 members come from some 40 countries and represent a wide variety of human experience. There are perhaps more scientists, both natural and social, in its membership than other categories, but it also has a number of industrialists, political figures, humanists, and philosophers. Members are invited by cooption of the Executive Committee, and this leads to some accusations of elitism. In practice, no other way seems to be possible if the informal nature of the Club, which its founders regarded as essential, is to be maintained. In reality, suggestions for membership come from many quarters and are by no means the prerogative of a few in the center. In any group of people, however carefully selected, sustained interest and work can never be assumed. A number of members of the Club, not having shown signs of activity in the past two years, were tactfully invited to withdraw and become Associates to make place for others. In this way it has been possible to achieve a better balance of nationality and of ideology than was represented by its initial membership, which was influenced by its European origin. Thus, there are now many individuals from the developing countries and a few, significantly increasing, from the Marxist countries.

The central concern of the Club of Rome is with what it calls "Le World Problematique," that tangle of interacting problems which we now face and within which it is increasingly difficult to formulate discrete issues and discrete solutions. The perspective of the Club is thus essentially global. It has two functions:

(1) Stimulation of research for the better understanding of the Problematique and its interactions (i.e., of the operation of the world system); and
(2) Bringing the results of its findings to the attention of decision makers in government, industry, and so forth, and holding dialogues with such people concerning the increasing importance of global problems and of longer-term thinking.

In fulfilling these functions, the Club restricts its activities to those of a catalytic nature, since with its absence of organization and structure, budget and manpower, it cannot hope to achieve substantial operational work.

THE EARLY HISTORY

As far as I am concerned, the need to create such a body germinated about 1967. I had been working for years on the problem of stimulating economic growth between the OECD countries, this having derived directly from the stress on industrial productivity and growth which followed the postwar reconstruction years. Perhaps we have been too successful, but by 1967, the growth situation seemed to me to be beginning to get out of hand. Disparity between the rich and poor countries was growing, and the unwanted side-effects of technological development on which growth depended were beginning to raise their ugly heads. Growth appeared to have become an end in itself and quality aspects were largely ignored. Within an organization such as OECD, despite strong concern on the part of its Secretary General at the time, Professor

Thorkil Kristensen, it was difficult to interest the member countries, all devoted to the growth mania, to look at the quality aspects and long-term consequences seriously, apart from problems of the environment which, as a result of popular protests, became very respectable a few years later. Furthermore, although OECD is professionally an extremely competent body, it appeared to me more and more that such an organization could not move sufficiently quickly to adapt to the rapid rates of contemporary change, and hence that there was need to reinforce the international organizations serving the national administrations by the creation of informal, complementary bodies unbound by party political restraints, not committed to immediate issues, and able to indulge in unorthodox thinking.

Toward the end of 1967, a letter arrived from Carroll Wilson of MIT, who enclosed a copy of a speech on long-term prospects for Latin America, seen in a geopolitical context, signed Aurelio Peccei, the Military Academy, Buenos Aires, and dated November 1967. Apparently, a copy of this lecture had fallen into the hands of one Jermen Gvishiani in New York. Gvishiani is professor of management at the University of Moscow, vice-chairman of the Ministerial Council on Technology, and, incidentally, son-in-law of Kosygin. He was extremely impressed by the lecture and wanted to contact Peccei, but had no idea who or where he was. He had therefore written to Wilson, his American colleague on the UN Committee for Science and Technology for Development, and Wilson, likewise not knowing Peccei, passed it on to me. I found out who Peccei was, wrote to him with the contact for Gvishiani, and suggested that we might meet when he was next in Paris. We had frequent meetings during the next couple of months, found we had common concerns, and felt that we had to do something about them.

Our chief preoccupation at the time was the apparent blindness of the Europeans to face up to complexity and change, and we therefore decided to call a meeting of European personalities to discuss these matters, and particularly the place of Europe in the world in a longer-term perspective. Peccei persuaded the Agnelli Foundation of Turin to finance the meeting and I lent Eric Jantsch, who was one of my consultants at the time, to do a background paper for it. The invitees were selected by Peccei and myself and included a range of scientists, industrialists, humanists, and bankers. It took place at the Academia dei Lincei in Rome, the oldest scientific academy in the world, and was a complete flop. The French, for example, quibbled on semantics and spent a large part of the meeting in explaining how the word "system" had no meaning in the French language; others took an anti-American stand; and at the end of the meeting, there was absolutely no consensus.

Nevertheless, after dinner that evening, in Peccei's apartment, a few of us decided that the matter was far too important to be left in abeyance, and that we should continue discussions in a small circle to try to formulate more clearly whether we should undertake studies or political action in some way. This discussion group we termed The Club of Rome because of the location of the first meeting. The name has proved somewhat ambiguous and there has been a tendency for some people to associate us with the Vatican, the Treaty of Rome of the EEC, or even with some of the features of Imperial Rome. Even among the initial six members there were disagreements at the second meeting, since two of them insisted that to look at problems on a world basis would be impossibly general, theoretical, and idealistic—we should concentrate on a specific problem such as urbanism—while the other four of us insisted on maintaining a global approach and looking at the interactions of the Problematique.

The four of us continued to meet frequently during the next eighteen months, generally in the Batelle Institute of Geneva, whose director, Hugo Thiemann, was one of us. We then began to coopt one or two others, and at the center of our discussions in Geneva was Hassan Ozbekhan, now of the University of Pennsylvania, whose wisdom and experience were enormously influential. As our discussions developed, we began to formulate a project entitled "The Predicament of Mankind," and we had a growing conviction that the world problems were developing so rapidly and were so complex that it was our duty to take some action. It was at this stage, therefore, that the present ideas of the functions and shapelessness of the Club began to evolve.

I had in my mind as a precedent the Lunar Society, which met in Great Britain at the end of the eighteenth century, completely outside the establishment and which, with its multidisciplinary mix of members, its future orientation and holistic approach, triggered off the Industrial Revolution.

During this period, Peccei and I attended a seminar given by Ozbekhan and Jantsch at the European Summer University in Alpbach, in Tyrol. Here we met a number of other personalities who became early members of the Club, including the Americans Detlev Bronk and Paul Weiss, as well as Edward Pestel from Germany. At dinner after this meeting, we had a talk with Doctor Krauss, Chancellor of the Austrian Republic at the time, who was fascinated by the ideas and invited a few of us to come to Vienna for a further discussion with him, his ministers, and some Austrian personalities. This we did, and it was an interesting indication to us of the possibility of wide political influence. Soon after this, Peccei and I also talked to many political figures throughout the world. We had an early meeting in Washington convened by the Swiss Ambassador, to which a number of people from the Nixon Administration came, but at that stage of his administration, we made no attempt to penetrate the White House. Immediately afterward, however, we had an important series of contacts in Ottawa, which included ministers and Mr. Trudeau, the Prime Minister. However, it very quickly became apparent to us that while political leaders were perfectly willing to listen to us with courtesy and even interest, our ideas and suggestions seemed too far off to have any policy significance. We therefore decided to concentrate for a time on stimulating studies of the world situation, while not ignoring political implications and possibilities.

LIMITS TO GROWTH

We had been looking around the world for academic groups possessing a technique for investigating complex systems, of which the world is not the simplest. Of the very few possibilities, the best seemed to be offered by the systems dynamics approach of Professor J. Forrester of MIT. He attended the first annual meeting of Club members (then about forty), which took place in Berne in 1971, and gave a detailed account of the possibilities of his methodology. He also issued an invitation to send some members to Cambridge for a seminar in which a simplified world model might be discussed and elaborated. This took place immediately afterward, and at the end of this seminar it was decided to back up Forrester in his construction of a world model in which a number of the major material variables could be included, their interactions investigated, and their time-projections studied. Financing for this was obtained from the Volkswagen Foundation of Germany, and Forrester constituted a team of young scientists from different countries around one of his assistants, Professor Dennis Meadows.

This group's work was published a year later in popular form, while the technical data was somewhat delayed in appearance. The appearance of the book, *Limits to Growth,* was certainly unorthodox, and some would say academically unethical, since the appearance of a scientific work without the full data is unusual, to say the least. This was deliberate. As we saw the work develop, we realized that it had tremendous political significance and also that it would be exceedingly controversial. Knowing also the ways of academics, we felt that, first of all, publication of the full scientific results would inevitably delay the matter longer than public interest would permit; and that the controversy would probably drag on for a very long time. We did, however, take the precautionary measure of circulating some 80 copies of the manuscript to economists, natural scientists, philosophers, and others. There was practically no reaction; that was to come later.

In designing the experiment three basic points were made, and these are repeated several times in the book:

First, the initial model of the world system will inevitably be tentative and incomplete. Only a few variables will be used, owing to the absence of quantifiable data, and, as social factors will not be possible to include in the initial attempt, no allowance can be made for changes of value. Even for the variables employed, data are often inadequate and unreliable.

Also, the model will initially take the form of a global aggregation on the basis of the variables to be included. This has the obvious limitation that the conclusion of the model cannot be applied uniformly in an essentially heterogenous world. The objective, therefore, is not to give specific policy guidance but to allow an overall reflection on the material conditions of the world. Later work will be required to disaggregate, deepen, extend to include other variables including those of a social nature, and, of course, to refine.

Finally, the dangers of extrapolation are fully appreciated, particularly when examining exponential phenomena, which so often flatten out. The model is thus not intended to be prophetic but to indicate what might result if trends and policies remain unchanged, so that changes can deliberately be produced to prevent the projected results. Its concern therefore is essentially prophylactic.

In practice, many of the reviewers of the book ignored (or did not read) these assumptions, but instead restated them as criticisms.

The book was immediately controversial, the main attack coming from traditional economists, from those who maintain that growth is an inbuilt characteristic of society, without which stagnation, decay and disruption would result; and from the natural scientists with faith that solutions to all the problems of the world will flow from the cornucopia of science and technology. The tone of the attack was often strongly emotional and even hysterical. It should be realized, however, that the respectable economists of a few countries such as Japan and the Netherlands—overpopulated, overurbanized and overindustrialized to an extent that the consequences of technological development could not be ignored—agreed, although with some reluctance, that the book's description of trends was valid.

In many of the industrialized countries, the concept of a "steady-state economy" which Meadows propounds in his final scenario, is synonymous with stagnation. This might be very well the case if zero growth were to be accepted in a black and white sense. It should be remembered, however, that the idea is by no means new, and that the answer to this fear was in fact given by John Stuart Mill (1857: 754): "It is scarcely necessary to remark that a stationary position of capital and population implies no stationary state of human improvement. There could be as much scope as ever for all kinds of mental culture and moral and social progress; as much room for improving the art of living and more likelihood of its being improved."

Certainly the first Club of Rome study had many imperfections. It is too pessimistic with regard to materials, its assumption of reserves five times the present being much too low. Furthermore, it ignores almost completely the influence of the price mechanism. With regard to aluminum, for example, supplies of bauxite are certainly limited, but if the cost of the metal were more than doubled, processes for its extraction from the limitless deposits of clay would probably become economic. Many of the possibilities of material substitution are, however, highly energy intensive and this underscores a fundamental gap in the model, namely, the absence of an energy parameter, for which data were available but which was not remarked on by any of the critics. The contributions of technology were certainly understated, but it would probably have been unwise to make any quantitative provision for this at the time.

The book had an immediate worldwide impact; apparently it gave substance to the intuitive feelings of many, although I doubt that it would have attracted more than passing interest had it appeared five years earlier. Over three million copies have been sold, in about thirty language editions ranging from Icelandic to Thai and Chinese. In some countries, such as England, sales were remarkably low, while in others, such as the Netherlands, fifty thousand copies were sold on the day of publication, and the total edition approached nearly half a million.

The book generated innumerable newspaper and magazine articles, radio interviews, and television features. In Italy, for example, there were three hour-long television programs on the subject, while the fairest and most skillful film was broadcast by the Canadian Broadcasting Corporation. Reaction in the Netherlands was quite exceptional. It led to the setting up of a group of prominent industrialists, bankers and politicians, who published a review of the national policy as it would have to be revised in the light of *Limits to Growth.* In Rotterdam, a permanent exhibition of the interconnections of the Problematique has been established. At its opening, the Queen of the Netherlands, in the presence of the Prime Minister and other Cabinet Ministers, presented copies of the Dutch version to boys and girls representing the various high schools of the region. The Club of Rome was awarded the German Peace Prize for 1973, to the accompaniment of nationwide broadcasts and debates.

Many discussions of the subject have taken place. Foremost among these was a large conference organized at UNESCO by Valéry Giscard d'Estaing, with many European leaders participating. Sicco Mansholt, then the president of the European Economic Commission, in a famous open letter to the Commission, espoused all too uncritically the cause of zero or even negative growth and this became a major issue at the last General Election in France. Many student meetings on the subject were held, and personally I have addressed thousands of students in four continents. One of the more touching occasions was at the University of Strasbourg where, as a prelude to debate on *Limits to Growth* in which I took part, the students performed a ballet which they had composed on the subject, with the variables interacting through the choreography. Open and closed meetings of industrialists, bankers and politicians on the book have taken place, not least in Japan. The members of the Executive Committee of the Club accepted an invitation from the Soviet government to take part in a seminar on *Limits to Growth* with a number of selected scientific and political leaders in Moscow. It is amusing to recollect that within a single week, *Limits to Growth* was discussed by the Board of Standard Oil of New Jersey, an assembly of the Gnomes of Zurich, and the Tupamara in Argentina. On the whole, however, there was great resistance to the ideas in the less developed countries, where there was great suspicion that the work was a neocolonialist apologia, the rich countries having concluded that in the pursuit of wealth they had fouled the nest and wanted to clean up, but in the process found it necessary to warn the developing countries to stop having so many children and not to expect too much because of a slackening of growth. It is significant, however, that in recent years, as the work of the Club developed, its obvious sympathy to the problems of the poor nations has reversed this attitude, and we are often now regarded as class traitors in the rich countries, too closely identified with the problems of economic disparity.

To sum up on *Limits to Growth,* it seems to me that it was extremely important for three reasons. First, it has triggered off a debate throughout the world which was extremely timely and necessary. This would not have happened had it not made concrete the disquiet of large numbers of thoughtful people and especially those in the highly industrialized and the highly populated countries. There was a danger of course that this debate would polarize extreme positions, some taking the apocalyptic view which can easily generate fear, despair, or resignation on the part of the young, in contrast to the Micawbers, who assume that everything can be left for science to turn up. The Club of Rome, of course, rejects both these extremes. The situation is well expressed by Phillippe de Seynes, at that time Under-Secretary at the United Nations for Economic and Social Affairs: "It is probably fair to say," he states, "that one of the striking developments of 1972 was the sudden popular awareness of the limits of our planet. This awakening, dramatized by sensational publications and passionate controversies, is little by little affecting the practical men in governing circles and can influence, in more than one way, first attitudes and then policies."

A second useful impact of *Limits to Growth* is a new awareness of the importance of the interaction of different trends and different sectoral policies. This has been greatly reinforced as a result of the petroleum crisis. Never again will it be possible to regard energy, population, agricultural production, monetary affairs, or industrial growth as autonomous areas of policy determination.

Finally, *Limits to Growth* is important as the first serious attempt to break into a new field of research, that of understanding the quantitative workings of the complex world system. Admittedly, it is a faltering and imperfect beginning, but the most contemptible criticisms of the work were those which denounced it for not having explored the whole of this vast new field of inquiry, and for not having given all the answers. As a pioneering effort it has stimulated scores of new interdisciplinary investigations in different parts of the world. Initially, some of these sought merely to evaluate or discredit Meadows' original study; others are attempting to deepen or extend it; still others are introducing new dimensions or trying out different methodologies in studying global regional or national situations.

REACTION OF THE CLUB MEMBERS

The Club was surprised by the impact of the first research project it sponsored and not a little embarrassed by its identification in the public mind with zero growth. In fact, we had scarcely discussed the topic before the book appeared. Such identification was not surprising, considering that the Club of Rome was at that time completely unknown, and that its first effort reached radical conclusions on growth. Newspaper comments on the meeting of the Club in Philadelphia stressed that the Club has changed its mind or changed its attitude. This is basically untrue. As a Club, we never espoused the cause of zero growth and, from the beginning, we had many other interests within the web of the Problematique. It is seldom recognized that *Limits to Growth* is a report to the Club and not by it. This is not to say that we are uninterested in growth, but perhaps as a consensus we prefer to stress problems of the quality rather than the quantity of growth.

At the first meeting of the Club after the publication of the book, held in Paris, many members voiced their disquiet at this identification with zero growth, although there was a consensus that the book was salutary and that discussion of the growth problem was highly necessary. In fact, we were all looking toward other kinds of growth. We felt it necessary to issue a statement as to the general aims and objectives of the Club, and this was made available immediately after the Paris meeting as a pamphlet entitled "The Club of Rome–The New Threshold." This was read into the Congressional Record in Washington by Senator Pell. We found it easiest to explain the Club's attitude by stating what it is not. It is:

> *Not* a group of advocates of zero growth, although we feel strongly that the nature, tendencies, qualities, and consequences of growth require deep and continued analysis and discussion.
>
> *Not* a Club devoted exclusively to the problems of the industrialized societies, attempting to find solutions to the difficulties of affluence, but a group concerned with the world situation as a whole and with the disparities it includes.
>
> *Not* a group of futurologists, but of individuals who realize the necessity of attacking the longer term and fundamental problems which are difficult to approach with our present methods of government and which could give rise to irreversible situations.
>
> *Not* a political organization, either of the Right or of the Left, but a free assembly of individuals seeking to find a more objective and comprehensive basis for policy-making.
>
> *Not* a body devoted to public propaganda for change, although, should we succeed in better delineation of the Problematique, we are convinced that our results should be made known universally through appropriate national and international organizations and media.

THE SECOND GENERATION OF REPORTS

Just after the publication of *Limits to Growth,* the Club held a meeting at Rio de Janeiro with a number of Latin American personalities, at which some of the attitudes of the developing countries toward its work, just mentioned, were brought out forcibly. We suggested that the Latin Americans should themselves undertake a study from the point of view of the underdeveloped countries. This was accepted, and the Bariloche Foundation of Argentina undertook the task, with an international advisory group and led by Amilcar Herrera. Funds for this were provided by the International Development Research Council of Canada at the suggestion of the Club. This extremely egalitarian model has recently appeared under the title "An Alternative World Model." It sticks rather closely to the Meadows methodology and examines the possibility of providing the basic necessities of life—food, housing, education, health care, and information—at a decent level, and as far as possible without external aid. The report is rather optimistic, except for the food difficulties of a swollen population in South East Asia.

A further study was initiated to meet the criticism that the first report had underestimated the importance of science and technology. We invited a group of "technological optimists" to tell us how research and development could help to solve the problems of three sectors—namely, energy, materials, and food and agriculture—for the increasing world population. This group, under the chairmanship of Dennis Gabor, did a very thorough investigation, consulting a wide range of experts. The study is about to appear under the title "Beyond the Age of Waste," under the signature of Gabor, Umberto Colombo, and myself. In fact, the scientific experts were optimistic within each individual sector, but greatly tended to underrate the interactions between the sectors; for example, the vast needs for increased energy to obtain widespread increase of agricultural yields through fertilizer productions, desalting of water, and so forth, were neglected as not the affair of the agriculturists, while environmental and climatic effects of cutting down jungles in order to grow rice, and so forth, were regarded as equally outside their scope.

More important is the study by Professors Mesarovic and Pestel in creating a disaggregated world model, using a methodology considerably different from that of Forrester. This work, undertaken at Case Western Reserve University and the Technical University of Hannover in Germany by a rather large multidisciplinary group of collaborators, has built a model in which the world is divided into ten mutually interacting and interdependent regions, each with a certain degree of ideological, economic, or environmental coherence. The regions are North America; Western Europe; Japan, Australia, New Zealand and South Africa; Eastern Europe including the USSR; Latin America; the Middle East; North Africa; Black Africa; South Asia; and China. The model does not assume the existence of any formal or informal supranational arrangements.

The technique operates through a causal stratum and a decision stratum, each of which is provided with a series of levels which vary in accordance with the type and complexity of the system being studied. For example, the causal stratum may have a natural level (involving geophysical, climatic, and ecological features), a technological level, population, economic and sociological factors. The main objective of this model is to provide the possibility of man-computer dialogue and interaction. It makes it possible for the decision maker or his analysts to work out the possible consequences of certain potential decisions or to provide alternative scenarios to assist in coming to decisions. It thus attempts to provide a policy advisory mechanism.

In the popular report by Mesarovic and Pestel (1974) entitled *Mankind at the Turning Point,* there is a discussion of a number of scenarios concentrating on a relatively small number of variables of particular significance: the population explosion with its demand for increased food supplies and energy; the gap between industrialized and developing countries; socioeconomic problems; depletion of nonrenewable resources. This model makes it possible to include in a computerized analysis of the consequences of alternative policies, social and political factors, in that these are inherent in the assumptions made with regard to the various policy alternatives.

This book, although many of its findings are quite dramatic, did not receive the widespread attention or attract the large sales as *Limits to Growth.* It does point out very great dangers with regard to the feeding of increased populations, particularly in Southeast Asia, and makes a very important point of the enormous cost in money and in human suffering if certain decisions are long delayed.

Public reception of this book again varied greatly from country to country. In Great Britain, for example, it was largely ignored or received somewhat contemptuous mention. In Germany, extensive summaries appeared in the main newspapers, while in France, thanks to a particular campaign mounted by J. J. Servan-Schreiber, it penetrated deeply. A weekly issue of *L'Express* was devoted to the subject, together with articles on the Club of Rome, portrait of Peccei, and so on. This achieved a sale of two million, and it is calculated that one person in four in France read it. More important is the present series of attempts to use the Pestel/Mesarovic method in long-term planning. The European Common Market organized a meeting at which a number of political figures from different European countries tried out the model with regard to energy policy. The consensus was that it was not yet sufficiently refined or sufficiently disaggregated to be of much use in day-to-day policy-making, but that it could be of great service in providing new insights on the consequences of long-term policies and in preventing counterintuitive errors which policy makers, with their relatively simple mental models, could hardly be expected to see.

The Mesarovic/Pestel model and its methodology is undergoing continuous development, and its utility in practice is being tested out in a number of real situations. It is being used, for example, in the long-term economic planning of Iran to give some indications of the consequences of alternative series of measures for the time when oil exports will have dwindled; it is being used in Egypt for studies of the relationships and potential developments between the various countries of the Middle East; it is being tested out in Germany in the context of inter-European connections and interactions with the Third World; in Venezuela, with the support of the President, it is being used to develop alternative strategies for the development of the South American continent. The initial findings of this latter operation were discussed in detail at a meeting of the Club of Rome in Caracas in 1977.

RESHAPING THE ECONOMIC ORDER

In 1975, at the instigation of Algeria, supported by a number of the uncommitted countries, an important debate took place at a special session of the General Assembly of the United Nations, on the need for a new international economic order. This was, of course, inspired by the sense of inequity and the gross economic disparities between the rich and the poor countries. In the course of the debate, it was obvious that the significance of the term "new economic order" was quite vaguely understood, and that, indeed, any new order would have to include social and political, as well as economic, elements. The Club of Rome, therefore, felt that it would be useful to give some substance to the term and help to delineate its possible ramifications by commissioning a review of the existing order of things. Aurelio Peccei, therefore, invited Jan Tinbergen, the Dutch Nobel Laureate in Economics, to assemble a team of experts to do this. Tinbergen accepted, and the Netherlands government generously made available the necessary funds. A team of part-time experts was constructed and began to examine the various elements of the situation, such as monetary policy, the international trade and credit systems, the operations of the World Bank, energy, industry, technology, the ocean, the role of the transnational corporations, and so on. For each subject, a small group was at work for eighteen months, of which at least one member was from an industrialized and one from a Third World country. Meetings of the full group took place frequently, and Tinbergen and his staff were responsible for the synthesis, with comments and criticisms from the full team. Initially, the project was entitled "Reviewing the International Order" (RIO), but as the work progressed and suggestions for change began to come forward, it was decided to give it a more constructive appeal, and the final and rather substantial report was published in October 1976 as *Reshaping the International Order* (see Tinbergen, 1976).

During the preparation of the report, a very large number of political, economic, and other experts visited the headquarters in Rotterdam, from all three political worlds. This has encouraged the Netherlands government and Professor Tinbergen to develop a continuous dialogue on the subject, and a RIO foundation has been created for this purpose.

GOALS FOR A GLOBAL SOCIETY

The initial studies of the Club of Rome have concentrated mainly on the material aspects of the world Problematique and on subjects such as the population explosion. This was frankly because these subjects are easier to comprehend, easier to quantify, and ideologically less controversial. Throughout our discussions, however, we have always been mindful of social and political aspects and have realized how deeply many of the problems involve the value system of society. We feel that the time has come to take the plunge and begin to look into some of these difficult questions. Ervin Laszlo was invited to gather together a team to investigate what could be goals for harmonious survival of the various societies of the world. This work is now in progress and the initial suggestions are under discussion with groups throughout the world representing the main ideologies and religious and cultural divisions. This important work is obviously only a beginning, since the subject is so vast and so important, and we expect to have to launch follow-up studies of various kinds.

Other projects under study by the Club of Rome at present include a study of world power structures, formal and informal, as well as one on the educational requirements for a new world, the latter on an initiative from Rumania.

MINISTERIAL MEETINGS

The Club attempts to hold annual meetings of all its members. In view of its loose structure and especially of its deliberate avoidance of financing obligations, this is somewhat difficult to arrange. In each case, it is necessary to find a host willing to provide the quite considerable support that is necessary. A series of such meetings has taken place in Bern, Canada, Paris, Tokyo, and Berlin, with occasional partial meetings such as that held in Philadelphia. The forthcoming meeting in Algiers will be in the series of annual meetings, to be followed by others in Caracas and Stockholm. On each of these occasions, in addition to a restricted gathering of the members, there are discussions of specific issues with a wider participation and usually smaller meetings between the members of the executive committee and political personalities of the country. For example, at the end of the Tokyo meeting a very useful discussion took place with ministers, and the Club received a scroll of appreciation from the Prime Minister.

Even in the earliest discussions of the Club there was mention of the eventual need to stimulate the creation of a corresponding apolitical and informal grouping of present and past senior political figures to discuss longer-term world problems, without briefs and unreported, as a means of establishing understanding of critical issues. This seemed, and indeed still does seem, a long way off. However, several years ago we felt that our intermittent and individual discussions with political leaders should be reinforced by collective discussion. We decided on a bold and, some would say, even impudent approach. The members of the executive committee as private citizens from so many countries invited a number of prime ministers and presidents of countries to meet with them for a couple of days of private discussion—unbriefed and unreported. Rather to our surprise, they said yes, although in fact it was impossible to collect them together on a single day owing to their other commitments.

First of all, as always, we had to find a host willing to pay for the meeting. Bruno Kreisky, the Chancellor of Austria, agreed to do this, and the meeting was held at Schloss Klessheim, in Salzburg. There was no intention of inviting senior politicians from the bigger countries who, had they accepted, would have dominated the discussions; but we explained our intentions beforehand in Washington and Moscow, to prevent misunderstanding of our intentions, and we had the benefit of cables wishing us well from the American Secretary of State and from Dr. Waldheim, the Secretary-General of the UN. Those attending were, in addition to Kreisky, Nello Cellio, lately chairman of the Swiss Federal Council, Presidents Senghor of Senegal and Echeverria of Mexico, Prime Ministers Trudeau of Canada, Olof Palme of Sweden, and Van der Ueyel of the Netherlands, as well as ministers representing the heads of the states of Ireland, Pakistan, and Algeria. The Australian minister was unable to reach Salzburg because of floods. Only seven members of the Club participated. We provided the chairman with brief papers on a number of issues. The dialogue was lively, with few major disagreements, and it gave the ministers a rare opportunity to discuss world issues without briefing from their officials, with complete frankness, and in a relaxed atmosphere which is not possible at traditional international meetings where negotiations take place or where national declarations of position are made for home consumption. If the trend of the Salzburg meeting could be stated in a single phrase, it would be: "The need for world solidarity in the face of global problems."

On the eve of the Salzburg meeting, Chancellor Kreisky gave a fireside talk to the Austrian people, in which he explained what the Club of Rome was and why he had felt it useful to host the meeting. He ended by saying: "This is why I shall be inaccessible during the next few days; I shall be putting Austria above party and the world above Austria."

It is not possible to repeat such a meeting frequently, so the next year when President Echeverria of Mexico suggested that he would support a similar meeting in his country, we did not invite the top ministers, but rather asked the various heads of state to delegate a minister to represent them. Ministers from 26 countries attended. The meeting was useful, but less fruitful than Salzburg; there were just too many people present to permit a real give-and-take dialogue.

THE NATIONAL ASSOCIATIONS FOR THE CLUB OF ROME

As already indicated, the small size and free-ranging nature of the Club and its determination to restrict its energies to strictly catalytic functions, which has been the secret of its vitality, have made it impossible for us to work in depth in any individual country. Nevertheless, in many countries our work has evoked considerable interest and a desire to follow up on the national level. All of us are under constant pressure to aid national activities and have devoted perhaps too much time to speechmaking. However, quite a number of national associations of the Club of Rome have sprung up spontaneously, without any stimulation from the center. These are in no way controlled by the Club, although they have generally formed around national members. They vary greatly in form, objective, and activity. The first was created in the Netherlands immediately after publication of *Limits to Growth.* The next association came in Japan, well financed and devoted mainly to research on various aspects of Japan's relationships to the rest of the world, and especially Southeast Asia and to the inclusion of social factors in comprehensive models. The Canadian Chapter is modelled after the main Club, with 100 members; it was given a good impetus by the Prime Minister. The Australian Association is rather like that in Canada, while that of Switzerland has a small selected membership, and is discussing deeply the need for a reorientation of national policies to a new kind of world. The group in Finland is studying the place of small, developed countries in a new type of world still dominated by the big powers. An American Chapter of the Club of Rome was created at the end of the Philadelphia meeting.

ADDENDUM ON IFIAS

The International Federation of Institutes of Advanced Study (IFIAS) is, in some ways, to be seen as a scientific extension of the Club of Rome. There is absolutely no formal link between the two bodies, but, in addition to the directors of the affiliated institutes, five out of the nine noninstitutional members of the Board of Trustees, as well as its General Secretary, are prominent Club personalities. IFIAS was created as a consequence of a symposium convened by the Nobel Foundation in Stockholm in 1969, which was attended by scientists (including a number of Nobel prize winners), humanists, and public figures, to discuss the role of science in relation to human values. It consists of some 23 member institutes, including the Pasteur Institute of Paris, the Neils Bohr Institute of Copenhagen, the Karolinska of Stockholm, the Institute of International Affairs of Geneva, the Walter and Eliza Hall Institute of Medical Research of Melbourne, Australia, Woods Hole, and NCAR of Boulder. The objective is to help the member institutes to develop their work in a socioeconomic perspective and to formulate and initiate multidisciplinary research projects on long-term world problems. Already some twelve projects are in operation; for example, on the effect of impending climatic change on agriculture and human settlement, loss of productive soil, energy analysis, and case studies of national energy futures, enzyme engineering, and optimum conditions for individual development. At its annual meeting of the Trustees at Aspen, Colorado, in 1975, there was an invitation to the chairman to attempt to write a report on "the state of the planet." This was published first in an English language version in London; agreements have already been signed for eight other language editions. This will be in two forms: first, the full version as a thin paperback for decision makers and the informed public; and, second, a popular version, profusely illustrated, and selling on the bookstalls for a modest sum.

IFIAS thus shares with the Club of Rome not only an unusual flexibility and capability of quick action, due to its relative lack of structure, but also a declared intention to influence political decisions. It has already had some minor influence. For instance, the declaration on impending climatic change has been read, if not yet heeded, in many external affairs ministries; a similar manifesto will shortly be issued on the need to respect cultural diversity and to work through it, in development planning. The IFIAS pioneering work on energy analysis, i.e., assessment of the energy content of different products and processes, has been followed up by the European Common Market, while its case study of the future energy alternatives for Denmark, which indicates how demand can be lowered considerably and without sacrifice of either economic vitality or life quality, is under consideration by the Danish government which, at the conclusion of the project, assimilated the team into its advisory mechanism. The IFIAS work on enzymes, which ranges over a wide field of application, has already had considerable influence on research priorities in India and the Soviet Union. The latter country collaborates exceedingly effectively in IFIAS work. Our recent seminar on the loss of productive soil in the world, held at Samarkand, has already had an influence on agricultural and other planning, and is expected to be followed up vigorously with the support of various UN agency and national funds.

SOME CONCLUSIONS

It is difficult to state categorically how deep the political influence of the Club of Rome has gone. It has undoubtedly been considerable and, indeed, very considerable in comparison with most other NGOs. However, little of it can be seen in terms of actions as direct consequences of the Club's initiatives. Most of it is indirect and gradual, through achievement of a gradual change in public thinking, of changes of concern toward those of quality. Its influence has also been exerted through changing attitudes on the part of political leaders. Its views have been repeatedly quoted; for example, by many Western politicians such as the present President of France. It has probably influenced the population policies of Mexico. Its views are treated with respect in many political circles; for example, an important meeting took place in Madrid, where, on the invitation of the Spanish authorities, the Club of Rome sent a dozen of its members and other personalities generally sympathetic to its objectives to hold discussions with a group of prominent Spaniards from government and academia as part of the process of opening up the thinking of that country to wider issues after the long years of Franco introversion. After a very substantial and intimate discussion with the young King and Queen, the Club members introduced a series of debates on key world issues, including a whole session devoted to the IFIAS "state of the planet" report.

In addition, the Club has continuous and friendly discussions with the heads of the international organizations, with the Secretary General of the UN and the heads of many of its agencies, with the Secretary General of OECD, and many other bodies. It has helped to give some reality to the discussions of the New International Economic Order, and helped to lift the debate from the merely economic to the larger socioeconomic perspective. It has raised the question of the quality of growth as a national objective. Its members, individually and in groups, have had innumerable discussions with leading industrialist and banker groups and with students in many countries. This is already a substantial record considering its short life.

One of the characteristics of the Club is that it feels itself to be an essentially temporary organization—or rather, nonorganization. At most of the meetings of its executive committee, the question is raised, "Has the time come yet for us to disappear, or can we still have some influence as a pinprick to complacency?" The lack of any urge for political power, for the creation of a new party ideology or for mere self-perpetuation, is one of its greatest strengths.

If I were to be asked my opinion of the potentialities of the international organizations, both intergovernmental and NGO, the following would be my answer, of course in terms of gross overgeneralization. Both types are necessary and should evolve in a complementary sense; both types are generally extremely inefficient. The main intergovernmental bodies, including the UN and its specialized agencies, do not work well because there is little political will to make them effective. They are obsessed too often by immediate issues; they are given too few clear mandates; and, as far as the specialized agencies are concerned, the vast variety of national situations, environments, levels of economic attainment and objectives, is such that it is very difficult to reach any real agreements based on common self-interest. Too often their projects are accepted because they are regarded by all as harmless, rather than evolved to be of common utility. The staffs of these organizations, often intelligent and devoted, become inbred, frustrated, and increasingly far from real problems; their bureaucracies are terrible, their mandates overlap, and there is continual infighting for power advantages. Staff recruited on a basis of national quotas rather than of the best man for the job makes for mediocrity. One would imagine that a system built on regional groupings would be more effective in making possible a greater degree of common self-interest, each more comprehensive in its coverage since the distinction between the existing special agencies, of essentially historical origin, becomes more and more confusing and meaningless in view of the interactions of the world Problematique. Bodies such as OECD, of limited membership, have probably been professionally most competent when they have been politically weak, and the lack of commitment to the most immediate issues has enabled the member countries to tolerate greater initiative on the part of their secretariats, who have thus been able to go more fundamentally into the problems which are their concern.

The NGOs are still more varied in their nature, scope and effectiveness; there are probably much too many of them, and most of them are financially impossibly feeble. At best, they can be effective in minor policy influence, although usually only for brief periods. Most of them suffer from the same difficulty of too-heavy structures as the intergovernmental bodies, especially where their governing bodies are highly representative of academies and other national bodies. Hence there is continual reference back home before decisions can be taken, which makes for very slow progress. However, in both types of organization, dynamic individuals, with strong motivation and a clear sense of objective, can achieve breakthrough results—but this is the exception.

What seems to be needed is a thorough analysis of the NGOs with regard to objectives, financing, and structure, to determine the conditions which make for effectiveness and especially to make possible some degree of policy impact.

Returning to the particular for a few final words, I feel that the world action which a small group of citizens has been able to mount through the Club of Rome during these last few years indicates the need and importance of informal groups who, lacking political ambition or political dogma, can act swiftly and catalytically, something which the large, intergovernmental organizations, with their heavy bureaucracies and political constraints, just cannot manage, but which is particularly necessary today, when change is so rapid. IFIAS is at present considering how a network of such bodies might be constituted by informal consultation and complementary action between a grouping of such bodies such as IFIAS itself, the Club of Rome, Pugwash, and the Third World Forum. It is the time to experiment.

NOTES

1. For more information on the Club of Rome, see King (1974) for a historical overview, and King (1979) for a more detailed account of the Club and its work.

REFERENCES

KING, A. (1979) "The Club of Rome: a case study of institutional innovation." Interdisciplinary Science Reviews 4, 1 (March).

--- (1974) Center Magazine 7, 5 (September/October).

MESAROVIC, M. and E. PESTEL (1974) Mankind at the Turning Point. New York: E. P. Dutton.

MILL, J. S. (1857) "Principles of political economy," in Collected Works.

TINBERGEN, J. (1976) Reshaping the International Order–the Rio Report. New York: Elsevier North-Holland.

COMMENTARY

Frederick W. Frey

Although social scientists all too often have their focal problems determined by policy makers and all too seldom select them as scientists, certain dramatic phenomena periodically occur that understandably command the attention of officials and scientists alike. Internationally, we confront the near swamping of the United Nations organization by the great influx of Third World countries in the past few decades, the population explosion, soaring energy costs, North-South imbalances, and so forth. Domestically, we witness escalating divorce rates, seemingly inescapable inflation, a probably marked future increase in the cohort of the aged, etc.

Though more modest in its implications, the recent emergence of thousands of international nongovernmental organizations (INGOs)—transnational actors impacting to an undetermined degree on both national and international activities—is another extremely significant development. By 1974, there were nearly 2500 such entities, more than a third of which were international scientific and professional associations (ISPAs). And the number has been growing at approximately 6 percent per year. (Merely listing them and appreciating their variety threatens the uninitiated with "acronymia"—a stupor induced by overexposure to unfamiliar acronyms.)

The four chapters of Part Three furnish diverse initial perspectives on the political impact of international scientific and professional associations. Anne Feraru looks at the role of ISPAs at important international conferences of the United Nations; Michael Pentz and Gillian Slovo examine the influence of one of the earliest and best known international scientific conferences—Pugwash; Edward Miles compares the impact of the Committee on Space Research (COSPAR) and the Scientific Committee on Oceanic Research (SCOR); and Alexander King describes the intriguing activities of possibly the most famous INGO of all, the Club of Rome. What can we learn from these diverse analyses?

While all these organizations are different and no "theory" of the significance of international nongovernmental organizations leaps to the eye, a general political scientist's reaction to these accounts might well display some gratification. Though dealing with rather new political entities, the findings that so far emerge in this area are far from unfamiliar. Thus, they meaningfully reinforce our inchoate understandings of the behavior of interest groups, lobbies, organizations, bureaucracies, and structures and processes of power. Let us flag some of the more salient illustrations.

First of all, the mere fact that attention is devoted to international *nongovernmental* organizations reflects the growing tendency in political analyses toward a more fruitful conception of what is "political" and what constitutes significant political actors. The old focus on nation-states and governments, though sometimes dressed in the more modern garb of "systems," "inputs," "outputs," and "functions," has long proven more an obstacle than an aid to political understanding. It has induced scholars virtually to ignore the private sector of the polity, with consequences as lamentable as if economists ignored the private sector of the economy. Students of international politics have belatedly but happily discovered such "transnational actors" as multinational corporations; the focus of the present studies on other important nongovernmental organizations continues this salutary trend.

It is quite interesting that the functions of international nongovernmental organizations are so similar to those of their national counterparts, despite obvious differences in the environing political structure. Such nongovernmental organizations tend to surface in the interstices of the formal structure, accommodating to it ("structural parallelism"), but offering complementary capacities. They flourish, as the authors note, because of their relative informality, flexibility, rapidity, political insulation, and special expertise, filling seemingly unavoidable gaps in the formal apparatus.

Every major organization has a need for what are sometimes called "orthogonal structures," that is, components which are organized along different lines and by different principles from those generally prevailing in the basic organization. Such orthogonal structures are, however, extremely difficult to maintain in the face of a dominant and different organizational bias. The nongovernmental organization, national or international, seems to provide a relatively promising vehicle for this purpose. NGOs seem disproportionately prominent when matters involving unusual lateral coordination, specialized multidisciplinary effort, and unorthodox cross-cutting problems (epitomized by the "world problematique") are of paramount concern.

Some of the manifest weaknesses and difficulties of nongovernmental organizations correspondingly arise from their interstitial character. The INGO continually tries to walk a delicate path between loss of autonomy, on one side, and loss of power, on the other. If the NGO gets too close to the government, it may gradually become a quasi-official service agency and sacrifice its independence. If the INGO remains too aloof from government, it may sacrifice so much access and information as to lose its influence. Indeed, many of the INGOs listed by Feraru would fall into one or the other category of ineffectiveness, especially the latter. The "authoritative appraiser" role which many ISPAs seek, as Miles notes, easily slides into that of the formal authority or policy maker itself or into that of the very peripheral observer.

Analyses of domestic interest groups have frequently suggested the importance of *issue/phase* distinctions. The power of such groups varies not only across issues but also across phases within issues. Professional groups and staffs have been portrayed as having much more impact in issue formulation and decisional implementation than over actual decision or choice among alternatives. According to the present authors, a similar pattern of power prevails internationally.

Another important and intriguing domestic political phenomenon in the United States is that of "tripartite alliances" and what Douglas Cater has called "subgovernments." These are informal coalitions of executive agencies, legislative committees, and nongovernmental interest groups sharing goals and viewpoints. The subgovernments of oil, sugar, and agriculture are oft-cited examples—perhaps also even that of defense. *Mutatis mutandis,* the triadic alliances of officials of international governmental organizations, officials of national governments, and representatives of nongovernmental organizations may be a significant subgovernmental counterpart at the international level. In general, the "bureaucratic politics" of the international arena seem strikingly similar to those analyzed by students of national political systems, suggesting some of the strength of the dynamics of bureaucracy regardless of contextual variation.

David Truman's classic notion of "access" (being in a structural position to be able realistically to attempt influence), so useful in fathoming domestic interest activity, also seems reemphasized in these international analyses. The critical roles of such "notables" as Russell, Pauling, Einstein, Eaton, and others in Pugwash; of Revelle, Wooster, and similar scientific advisers in SCOR and COSPAR; and of King, Peccei, Thiemann, and others in the Club of Rome in providing access at the highest ministerial, academic, and industrial levels for their organizations, seem to account in no small degree for their relative success. For INGOs, "contacts" seem to be the name of the game, and at least this far they seem to be established more through judicious recruitment of members than through other avenues, though this should change over time. At present, the international prestige of some of its members may carry an INGO a good deal farther than any other single ingredient. With such prestige and access to funds, significant impact is apparently quite possible.

At the same time, conspicuous by their absence in the international arena (not surprisingly) are the opposite types of interest groups—those whose power comes less from the prestige and expertise of its members and more from their sheer number and aggregate occupational or social position (such as trade unions, veterans organizations). On the international scene today, externally-oriented, elite nongovernmental organizations abound; internally-oriented, mass nongovernmental organizations are not yet conspicuous. Obviously, the establishment and coordination of the latter type of INGO poses special difficulties. Nevertheless, one might expect them eventually to become more prominent, and when they do, the nature of international lobbying can correspondingly be expected to change.

Very little is said in these essays—understandably, given their pioneering character—about the significance of the private motives and personalities of the ISPA leaders or about salient cultural differences. Occasional reflections of such considerations do surface, however, as in Miles' comments concerning scientists' tendencies to employ their international participation as a tool for obtaining support for their research from their own governments, in his comments concerning the Weisman-Deacon conflict, or in his mention of the tension between physical oceanographers and fisheries scientists in SCOR. King's observations about the advantage secured by the Club of Rome from its clear lack of an urge for durable power is another intriguing suggestion that beckons for systematic analysis. Culturally, one perceives fascinating currents behind a number of the verbatim responses by Pugwash participants and glimpses the significance of cultural factors very explicitly in the French response at the initial Club of Rome meeting described by King. One hopes that such cultural dimensions may be more systematically explored in future analyses, since they seem so strongly characteristic of international, as opposed to intranational, interaction.

In sum, then, from these four glimpses of international scientific and professional organizations we obtain some tantalizing insights into a new breed of political organization. That this new breed seem so far to behave in generally understandable fashion is more significant. If true, and not merely the result of unrecognized parochialism, it markedly increases confidence in our inchoate organizational theories, importantly extending the range of variation in the types of units covered without exposing a need for major revision.

COMMENTARY

Robert S. Jordan

There has been increasing interest in international scientific and professional associations (ISPAs) and their role at the international level, because of their participation in the nongovernmental forums and "counterconferences" that have accompanied each of the recent major ad hoc global conferences organized by the United Nations. While much publicity was accorded to the unofficial forums that took place coincident with the United Nations conferences on the environment, population, food, status of women, human settlements, and trade and development, little is known about the role and contribution of ISPAs in the resolution of problems of international concern.

According to Anne Feraru, the 1974 *Yearbook of International Organizations* lists 901 ISPAs in the following categories: social sciences; law, administration; professions, employers; economics, finance; technology; science; health, medicine. At the Stockholm conference on the environment, 73 had observer status. The numbers varied from conference to conference, with 36 having observer status at the Vancouver conference on human settlements. These numbers refer solely to ISPAs and not to nongovernmental organizations (NGOs) in general.

ISPAs range from relatively loosely structured associations such as the Pugwash group (which has held 24 conferences and 25 symposia, or a total of 49 meetings, in 20 years) to complex associations such as the Committee on Space Research (COSPAR).[1] COSPAR is a member of the International Council of Scientific Unions (ICSU) and together with the International Union of Radio Sciences and the International Astronomical Union forms the Inter-Union Committee on Frequency Allocation. COSPAR's membership consists of 35 nations and 11 international scientific unions. It has a president, two vice-presidents, a bureau, an executive council, a small secretariat, seven working groups, and several special bodies such as the Consultative Group on Potentially Harmful Effects of Space Experiments, which itself has several subgroups. Within this range of

ISPAs fall the more traditional scientific unions which are grouped in a federation under a council. The purposes and methods of ISPAs vary accordingly.

Edward Miles has pointed out that among other things ISPAs, particularly those that are members of ICSU, are forums for organizing and maintaining international contact among their membership, particularly with a view to the presentation and stimulation of research and the diffusion of findings; they are forums for the planning and organization of cooperative international research projects such as the International Geophysical Year (IGY); they are centers for the aggregation and articulation of interests in particular disciplines, and as such lobby at both the national and international level; they are used by the international community as repositories of highly technical advice; and they seek comparability of data within particular fields. They are also forums in which compromise solutions to intractable international problems can be explored, and in which mutual understanding can be developed.

There are a number of ways in which individual members of ISPAs can have an effect on policy on questions of international concern. At the national level, as lobbyists, expert consultants, members of advisory committees, or staff of government agencies, they can influence the development of national policy on international questions. At the level of national representation in international organizations, as expert advisers to official delegations, as members of delegations, or as lobbyists they can have an influence on the articulation and implementation of national policy and in the adjustment of national policy in the process of transnational negotiation. At the international level, as technical advisers or consultants to international secretariats, as lobbyists, or as members of international secretariats, they can have an influence on the articulation and implementation of internationally agreed policies.

Although ISPAs are nongovernmental, as is evident from the foregoing, they are not necessarily entirely free from the influence government can exert on them, inasmuch as their members are scientists and professionals drawn from government as well as from academia, industry, and the business world including transnational enterprises. In general, this intermix of governmental with the nongovernmental is considered advantageous as it permits access to governments with a view to making known the ISPA's ideas which are themselves the product of transnational interchange. From the point of view of governments, ISPAs bring together scientists and professionals of different nationalities in an atmosphere that Arnold Frutkin (1965) has said "minimizes political considerations, yet permits responsible and authoritative planning."

The constructive role of ISPAs is thus seen as aggregation and articulation of the ideas of scientists and professionals from different parts of the world, the ability to mobilize their members on any issue which appears important to them, and informing and educating governments about such issues. The Pugwash group has stressed the responsibilities of its participants to contribute directly to the solution of international problems, to make clear to scientists the social implications of their work, and to educate the public. Edward Miles has suggested that the governing bodies of certain ISPAs can function as coherent transnational delegations on issues of interest to them. He points out that in order for ISPAs to be effective there must be a unified viewpoint within the ISPA on a given subject, and the members must have access to the centers of government decision-making. However, Pugwash participants are reported as believing that their expertise in the subject under discussion is more important than their access to government. A survey of Pugwash participants by Pentz and Slovo produced the interesting finding that the participants are more confident of the effect of their conferences on international scientific cooperation than on the positions of their respective governments on issues such as disarmament.

Where issues are controversial, involvement entails high risks for the ISPA and success is a matter of skill. One of the determinants of success is quality of leadership. In this respect, it is interesting to note that in a setting and an era of increasing complexity, it has often been one man or relatively few men who have acted as the driving force (Roger Revelle for oceanographic research, Bertrand Russell and Albert Einstein for Pugwash), as a catalyst (such as Cyrus Eaton was for the Pugwash group), or as a transmitter of ideas (e.g., individual Pugwash conferees who influenced members of the legislative and executive branches of their respective governments). Each of the individuals who are members of ISPAs also have a multiplicity of personal and organizational links.

Other important variables are levels of funding, changes in the internal or external environment of the ISPA, and culturally determined differences in role perception. The viability of ISPAs, or at least some of them, appears to be related to the actual and potential demand for their services.

ISPAs are in a continual state of flux. As older problems are solved or overtaken by events, some people question the continued usefulness of ISPAs formed to aid in the solution of those problems, e.g., the Pugwash group that sought to provide a neutral forum for scientists from the United States and USSR. In a sense, when such ISPAs effectively carry out the tasks they set themselves (however difficult this may be to demonstrate conclusively), they work themselves out of business.

Alexander King, a co-founder of the Club of Rome, has noted the special relevance to this question to the Club's activities and deliberations. He argues that the Club of Rome has had considerable (though mostly indirect) political influence, especially in comparison with other nongovernmental organizations. However, he goes on to note that at most meetings of the Club's executive committee the question is raised, "Has the time come yet for us to disappear or can we still have some influence as a pinprick to complacency?"

The Club of Rome is particularly interesting in this regard, because it was initially designed as an essentially temporary organization. King explains that there is no formal secretariat, nor is there a budget. The executive committee is composed of only eight people who meet rather frequently. The Club seeks to serve as a catalyst, stimulating research in selected areas and bringing the research findings to the attention of decision makers. The dilemmas involved in determining when to terminate the activities of an ISPA are real. Should one proceed to new, perhaps less appropriate, problem areas as the organization's functions in older ones are completed?

Another consequence of the activities of at least certain ISPAs is that they open new channels for communication, information exchange, and expert evaluation, which then exist independently. As new problems gain recognition at the international level, new ISPAs are formed. This took place, for example, as interest increased in deep sea research and in the implications of such research. As a result the Scientific Committee on Oceanic Research (SCOR) was formed.

The experience of SCOR is interesting inasmuch as its concerns are relevant to the ongoing Third United Nations Conference on the Law of the Sea that opened its seventh session in March 1978. The first issues with which SCOR concerned itself when it was established in 1957 were the use of the deep sea for waste disposal, the ocean as a source of protein, and the role of the ocean in climate change. SCOR has been closely associated with UNESCO and its affiliated intergovernmental Oceanographic Commission (IOC), both of which request advice from SCOR, thus providing a closely linked governmental/intergovernmental/nongovernmental arrangement. However, the developing countries in the Law of the Sea Conference have been resistant to the wishes of IOC and SCOR for a significant expansion of oceanographic research. It is thought that this attitude may have been caused in part by the fact that countries with large-scale maritime research programs have failed to provide significant resources for technical assistance, training, and education in maritime science in the developing countries. In addition, most members of SCOR, with the exception of the U.S. national committee and U.S. scientists, were reluctant to become involved in preparations for the Law of the Sea Conference and did not accept various suggestions to this end, including a proposal that SCOR should offer its services to the United Nations Secretariat for the conference in order to provide scientific advice on issues affecting marine science. The experience of SCOR is therefore something of an object lesson to ISPAs that may have a role to play at the international level in the future.

Some ISPAs have consultative status with the United Nations Economic and Social Council; some do not. Provision has also been made for consultative status with most of the specialized agencies of the UN system.[2] In addition, there have been specific resolutions and other efforts to insure that those in this consultative status are in a position to contribute constructively to the preparation and proceedings of various UN conferences.

ISPAs played a role in connection with the United Nations conference on water and desertification that took place in 1977; they will have a role to play in the activities of the United Nations Decade for Women (1976-1985) as it gets under way; and their cooperation was sought in the preparations for the 1979 United Nations Conference on Science and Technology for Development. It is in the preparatory phases of such conferences that ISPAs can exert influence because international conferences are of short duration and delegations are extremely busy throughout the conferences. ISPA influence in the implementation of decisions is also likely to be greater than their influence at the conference itself.

A new development has been the effort by the Secretariat of the UN Conference on Technical Cooperation Among Developing Countries (TCDC) to ensure participation by Professional, Technical and Voluntary Associations (PTVAs) from the developing countries in the conference, that was scheduled for August of 1978. The distinct terminology is being used to separate PTVAs from NGOs and ISPAs, which are commonly based and centered in developed countries. Thus far there are plans for holding a parallel conference of PTVAs and for some preparation of discussion papers in connection with TCDC.

A number of ISPAs have continuing relationships with United Nations Specialized Agencies. SCOR, for example, does work under contract for UNESCO and its IOC, and the Advisory Committee for Marine Resources Research does work for FAO. ICSU's Scientific Committee on Problems of the Environment (SCOPE) and the International Union for the Conservation of Nature and Natural Resources have close working relationships with the UNEP Secretariat and perform contract services for UNEP programs such as Earthwatch and the International Referral System which, respectively, conduct environmental research and evaluation, and information exchange.

Indeed, it is the ISPAs in the "hard" sciences that seem to have played the greatest role at the international level thus far, although others have also been active. It was at an international meeting of biologists at Asilomar, California, in February 1975 that warnings were sounded that led to stringent precautions in connection with recombinant-DNA research. It may be noted that the scientific and technically oriented ISPAs also appear to have shown the most interest in international conferences. Of the 85 ISPAs that sent observers to either the UN environment or human settlements conferences, or to both, 19 were scientific and 15 technological associations. In each of the other categories of ISPAs, not more than 13 associations were represented. It is particularly interesting that, as reported by Feraru, only two organizations that were classified as economic or financial ISPAs (the European League for Economic Cooperation and the International Union of Building Societies and Savings Associations) sent observers to Stockholm, and only one of them sent an observer to Vancouver. There would seem to be a greater role for ISPAs representing disciplines other than the "hard" sciences. However, it will not be easy, as the experience of SCOR with the Law of the Sea Conference has shown. (In this connection it may be worth noting that it was ISPAs with specialized expertise that were particularly helpful in maintaining some East-West dialogue during the Cold War.)

The problem areas of the immediate future are likely to be on questions related to the so-called "North-South" negotiations over the establishment of a new international economic order based on equity, social justice, and meeting the requirements of the most needy. If, as seems generally agreed, some ISPAs, such as the Pugwash group and COSPAR, were extremely useful points of contact between the Superpowers at a time of political tension between the two, do ISPAs have the same potential in the "North-South" dialogue between the industrially developed and the developing countries?

The editor of the *Bulletin of the Atomic Scientists,* Eugene Rabinowitch (1965: 14), once said "The Pugwash 'movement' is much weaker than some people outside may think—but it is the only international movement trying to change national opinion and political leadership everywhere toward better understanding of the challenges of the scientific revolution and of policies adequate to cope with these challenges." Is there, in the world of today, a similar role for economic and social scientists through ISPAs, particularly those whose membership includes persons from developing countries?

NOTES

1. It is acknowledged that some would argue that the Pugwash group does not strictly constitute an ISPA. However, its meetings of scientists must be considered at least as a closely related phenomenon, deserving of some commentary here.

2. For an extensive explanation of these arrangements for consultation within the UN system, see Andemichael (forthcoming); also see ECOSOC Resolution 1296 (XLIV).

REFERENCES

ANDEMICHAEL, B. (forthcoming) "Cooperation between nongovernmental organizations and United Nations agencies," in B. Andemichael (ed.) Non-Governmental Organizations in International Co-operation for Development. New York: UNITAR.

FRUTKIN, A. (1965) Space and the International Cooperation Year. Washington, DC: Government Printing Office.

RABINOWITCH, E. (1965) Bulletin of Atomic Scientists 21, 4: 9-15.

PART IV:
SOCIAL AND CULTURAL IMPACT OF ISPAs

11

DO ISPAs PROMOTE GLOBAL INTEGRATION?

Robert C. Angell

Since the middle of the nineteenth century, there has been a steady growth of both international scientific associations, like those of the astronomers and the chemists, and professional associations, like those of the physicians and the engineers. Such associations are usually federations of national or regional bodies, but they may also be associations of individual members. Though these international scientific and professional associations (ISPAs) have broadly similar purposes, they differ somewhat in general function and greatly in organizational detail. Since they have not been widely studied, any attempt to estimate their social and cultural effects can give rise to only tentative conclusions.

What international associations deserve to be called scientific or professional? There is clarity about the scientific ones. They are composed of members highly trained in one of the fields of knowledge who are trying to add to that knowledge through research. Doubts about inclusion in the category occur chiefly over whether the preoccupation of the members is science or pseudo-science. Most scientists would probably accept an association of parapsychologists but reject an association of astrologers. There is more difficulty with the professions. They are rendering services to the public, but so are many occupations that are generally not honored with the label professional. My inclination has been to include only associations whose members perform services requiring high intellectual or artistic competence. Doctors, lawyers, ministers and priests, engineers, architects, teachers, social workers, city planners, authors, accountants and auditors, librarians, and registered nurses seem to qualify. Some members of other vocational associations qualify too, but not all: associations of musicians, actors, visual artists, and businessmen, for instance. In such cases, one would have to look at the membership of an association to determine whether as a whole it was of professional caliber. An association of professional athletes would rarely qualify. I have also omitted travel organizations as primarily profit-making, not professional.

Organizations that carry out services closely related to a professional field, but in which other personnel play significant roles and have strong interests, have also been excluded. The International Library Association is not included because libraries are set up not by librarians but by public or private bodies that are interested in their conduct. Librarians are the dominant service-giving group, but they are not the policy makers. The International Hospital Association is similar. In this case, although doctors are often the controlling element, other trained personnel contribute significantly—nurses, administrators with business training, accountants, and legal counsel. Thus, the hospital does not represent the orientation of a single profession. The same is true of such bodies as the International Committee of the Red Cross and the League of Red Cross Societies. The World Council of Churches represents the same situation in another field. The fact seems to be that service-giving institutions necessarily develop a structure that embodies more than the ends and values of a single profession.

Both international scientific associations and international professional associations, then, are fellowships of high competence acquired by talent or training or both. The difference between the two sets lies in the objects of their endeavors. The scientists are bent upon adding to the corpus of knowledge in a discipline. Their achievements are registered in symbols expressing analyzed data. Their associations are devoted to communicating scientific findings, fostering new research, improving methods of inquiry and analysis, evaluating one another's work, and training programs for successors. The professionals are engaged in serving the public with the aid of a body of systematic knowledge, scientific and nonscientific. Their knowledge is put to work in practical situations, so that their achievements are registered in the lives of individuals, groups, and institutions. Their associations are concerned not only with competent performance, but with maintenance of ethical standards and the training of future professionals.

As one would expect, there are associations that are both scientific and professional. A good example is the Research Group for European Migration Problems (REMP). This was founded in 1952 by 36 Dutch, English, and German scholars, 28 of them in emigrant countries, 8 in immigrant countries. Members met in many conferences on European migration problems. In the first 14 years of operation, 28 studies were completed by members of the group and 8 by younger colleagues. The group brought up to date the study by the Population Division of the United Nations entitled "Economic Characteristics of Overseas Migrants, 1918-1954." It has also done studies for the Dutch government and the Organization for Economic Cooperation and Development (Beijer, 1972). REMP is a rather rare hybrid—an ISPA of purely scientific character focused on a pressing practical problem.

So much for the associations with which we are concerned. What kinds of effects are under study? Since national scientific and professional associations have usually preceded the international ones and have had more constant impact within countries, I shall mainly consider the effects of ISPAs on the world scene. More specifically, I shall be asking whether these associations promote intersocietal and intercultural integration.

By intersocietal integration I mean a process in which persons and groups in different societies develop cross-national bonds that satisfy their needs and interests, thus making for greater cooperation between societies as wholes. If intersocietal integration were occurring generally around the world, the prospect for development of a global society would be enhanced. There are many processes that are expressive of such integration. Among them are growing participation in international nongovernmental organizations (INGOs) and intergovernmental organizations (IGOs), more frequent conferences on world problems, increase in visits to relatives and friends abroad, growing amounts of assistance to developing countries, and more students and professors sojourning abroad for study, teaching, and research (Angell, 1969).

We must be careful, however, not to fall prey to circular reasoning here. Since part of the growing participation in INGOs occurs in ISPAs, and since ISPAs are seen here as causal to intersocietal integration, I have not included growth in INGOs as one of the effects of intersocietal integration.

By intercultural integration I mean the process by which shared symbols, beliefs, and values emerge from the contact and interaction of different cultures. This process is evidenced by linguistic borrowings between cultures, translations of books and articles, convergence of artistic tastes, sharing of religious beliefs and moral and legal standards, and greater consensus on the nature of world problems.

There are no studies that exactly focus on the problem being examined here. A paper by Louis Kriesberg (1972) entitled "International Nongovernmental Organizations and Transnational Integration" has an apposite effect variable, but the causal variable is too inclusive. There are many INGOs that are not ISPAs. Two pieces of mine—a chapter entitled "International Non-governmental Organizations" in *Peace on the March* (Angell, 1969) and the paper "National Support for World Order: a Research Report" (Angell, 1973) also focus on INGOs as they relate to international integration. I considered reworking the data of the second paper to include only ISPAs, but refrained because most of the components of the index used as the effect variable are political, not social or cultural.

In preparing this study, therefore, I have drawn the data almost exclusively from historical accounts of two kinds—those written about the activities of ISPAs over the years, and those written about intergovernmental bodies in which impacts on them flowing from ISPAs are discussed. This second source is inadequate, because the approach from the effect variable side—effect-to-cause rather than cause-to-effect—should include instances of ISPA impact on transnational intercourse and culture of all kinds, not just impacts on intergovernmental organizations. I have found a few indications of impacts of ISPAs on study and research abroad and a good deal on stimulating the growth in conferences on world problems, but such evidence is so fragmentary as to defy summary here.

Before proceeding to the data used, I wish to point out that historical accounts are most useful when they record the setting of an ISPA goal of promoting intersocietal or intercultural integration. One then has the clear task of determining the degree of success in attaining that goal. When specific goals are not recorded or when integrative consequences are unintended, it is difficult to ferret them out.

As an introduction to the analysis, I shall summarize the case histories of two ISPAs founded in the nineteenth century, so as to give some flavor of the types of interests expressed through these associations and the changing types of problems they encounter. The kinds of influence apparent in these two cases will be discussed in detail later.

THE INTERNATIONAL STATISTICAL INSTITUTE

The International Statistical Institute was founded in 1885, 34 years after the International Statistical Congresses began in 1851 (Nixon, 1966). These Congresses were quasi-IGOs, since governments named the representatives but did not give them instructions. Some of those attending were trained statisticians, others were not. The Congresses tried, without universal success, to get governments to organize central statistical bureaus and to agree on comparable statistical categories. The last Congress was held in 1878.

At the celebration of the fiftieth anniversary of the Statistical Society of London in 1885, the participants decided to found a private international organization in the field of statistics, its membership not to exceed 150 leading statisticians in various countries, new members to be selected by cooptation. The first Conference was held in Rome in 1887.

The Institute's history can be divided into three periods, 1887-1911, 1911-1947, and 1947 to date. During the first two periods, the Institute had the characteristics of both a professional and a scientific association. Its basic aims were to help governments attain statistical competence (in return for which the Institute received subventions) and to further the development of statistical theory and method.

The difference between the first and second periods was that what started as a fellowship of scholars and administrators became in 1911 a body with a bureaucratic nucleus. Policy remained under the control of members, however.

Many of the original members were chief statistical officers of European governments; and they continued the orientations of the preceding Congresses, fostering comparability of categories and building national statistical bureaus. Additional interests in stimulating the taking of censuses and in studies of population movements developed early. The improvement of trade statistics during the 1890s was largely due to biennial Institute reports on that subject. It is interesting that sampling was first discussed at the 1895 Conference. In the same period, Bertillon made two important contributions: developing a model list of causes of death, and proposing the classification of occupations (a suggestion carried out by the French government in 1909). Another field in which the Institute did pioneering work before 1911 was economic, especially trade, statistics.

At the 1911 meeting of the Institute, it decided to establish a Permanent Office at The Hague to conserve statistics of governments, to facilitate unification of methods, and to publish an international *Annuaire* and a bulletin. The Office was well supported by governments. The work of the Institute was interrupted by World War I between 1913 and 1923, but the Office carried on and produced five volumes of the *Annuaire International de Statistique.* In one of these, the Office proposed to governments the use of a standardized age distribution to compute mortality rates, a suggestion widely adopted. After the war, this series was carried on until World War II. In 1933, the scholarly journal *Revue Internationale de l'Institut de Statistique* began publication.

After World War I most members wanted the Institute to become the demographic arm of the League of Nations, though remaining an INGO. This never came to pass; but the Institute worked hard on labor and economic statistics, and its drafts in both fields were influential on the conventions adopted in 1928 and 1929. A mixed commission of League and Institute experts revised Bertillon's lists of causes of death in 1929. The Institute was moderately successful in dealing with the International Labor Office during the interwar period. The latter took over Bertillon's scheme of classifying occupations and refined it; and in 1932 called an International Conference on Migration Statistics at which many suggestions originating with the Institute were recommended by the ILO. On the research side, a committee report on sampling to the 1925 Conference of the Institute was authoritative; and its call for scientific sampling had wide influence among statisticians everywhere.

The setting up by the United Nations of a Statistical Commission in 1947 inaugurated the Institute's third period. The Institute was immediately given consultative status by the Economic and Social Council. That Fall, the UN convened a World Statistical Congress during which conferences were held on the future of the Institute. As a result, the role of the Institute changed: It became the world nongovernmental body at the heart of a network of national and regional statistical associations which had grown up through the years since 1887. Its functions were to upgrade statistics throughout the world by meetings, improved education, fellowships, advice to governments, furthering statistical comparability, and research. Thus, in the new phase, the Institute became an international scientific association of a more normal type, the chief abnormality remaining election of its members. Provision was made for ex-officio members (up to 150), which relieved somewhat the limitation to 300 regular members. Another traditional but unusual element remaining was that Conferences were invited by governments and the local expenses subsidized by them.

The third period has been the most creative one in the Institute's history. There has been new effort to improve industrial statistics. A committee publishes the *International Journal of Abstracts on Statistical Methods in Industry*. Perhaps the greatest emphasis has been on training statisticians worldwide. Under a UNESCO subvention, the Institute has set up regional units in Calcutta and Beirut for this purpose. There has been much activity in providing bibliographies and dictionaries of statistical terms. The Institute publishes the *International Journal of Abstracts: Statistical Theories and Methods*. The *Revue* continues publication, but now in English. Another line of effort is municipal statistics. This goes back to the Congress of 1872. Several volumes have been published on large European towns, and the members involved have become a special Section of the Institute. Because population research has mushroomed since World War II under the aegis of the International Union for the Scientific Study of Population, the Institute has established close relations with it. One example is the five-year World Fertility Survey, beginning in 1972, in which the UN Demographic Commission is also collaborating.

At the end of 1973, there were 609 members of the Institute from 80 countries. Ten international societies and 19 national statistical associations were affiliated.

The ISI's evolution has been most unusual. It started as a quasi-governmental body, though not one dependent on governments for decisions. It became an independent body of scholars devoted to the development of statistical science which concurrently turned over to UN organizations many of its earlier tasks. This reverses a frequent evolution from an INGO to an IGO—from the International Telegraph Union to the International Telecommunications Union, for instance. It is of interest that the committee that recommended the change in Institute function was headed by Stuart Rice of the University of Pennsylvania.

THE INTERNATIONAL COUNCIL OF NURSES

Unlike the International Statistical Institute, the second ISPA whose history I will outline was founded as the projected capstone of existing national associations. The International Council of Nurses (ICN) was conceived at the International Congress of Women in 1899 by Mrs. Bedford Fenwick of Great Britain, a nurse of wide administrative experience. She became the first president. The first Conference was held in Buffalo in 1901. The pattern of meetings was unusual—a Congress every five years and a Conference in collaboration with a national association every three years (Bridges, 1967).

The main effort from the first was to obtain adequate education for nurses worldwide. All states were urged to regulate the training of nurses by state examination and registration. At the first Quinquennial Congress in Berlin in 1904, the nurses of the United States, Britain, and Germany joined the Council as corporate associations. More nursing associations became affiliated at each Congress—19 by 1929, when there were 6200 registrants at Montreal from 38 countries. The number of affiliated national associations is now more than 60. An interesting reflection of the Council's origin is that for many years the Congresses passed resolutions favoring women's suffrage.

Among the notable changes that show increasing influence of the Council are the following:

- At Cologne in 1912, the Council established a Chair in Nursing and Health in honor of Florence Nightingale and sought funds to endow it. Later, this became the Florence Nightingale International Foundation.

- At Helsinki in 1925, the Council decided on a three-year course as standard for an association to obtain membership in ICN. At the same time, a permanent headquarters was established in Geneva. From that date on, the Council worked closely with the Red Cross and the ILO. In 1929, the ICN furnished information on unemployment among nurses to the ILO at its request.
- The Paris Congress of 1933 established a program of international exchanges among nurses.

The question of how the ICN should deal with unionization among auxiliary nursing personnel arose in 1948. That same year, ECOSOC and WHO granted the Council consultative status. In 1949, the Council's Executive Secretary addressed the Second World Health Assembly and was invited to join the First Expert Committee of WHO the following year. At the same time, WHO subsidized a study of advanced programs in nursing education to be carried out by the Florence Nightingale Foundation.

In 1950, the ICN also took over from the International Refugee Organization the Displaced Persons (Nurses) Register. The Council was more successful than were the physicians in placing refugees in new jobs.

In 1953, the World Federation of Mental Health obtained the cooperation of the Council and the World Hospital Federation in a joint study of mental health problems in general hospitals.

During the 1950s, the Executive Secretary of the Council visited many developing countries to advise national associations, governments, hospitals, and educational institutions. By 1957, the Florence Nightingale Foundation had become the central world authority on nursing education and curricula. About the same time, the ILO asked ICN to apply for consultative status. The Council then helped the ILO carry out the nursing part of a large study of professional occupations. The *International Nursing Review* appeared sporadically from 1930 to 1954; thereafter, it has been published continuously.

CHANNELS OF INFLUENCE OF ISPAs

So much for the two case histories. They show the great variety of influences of ISPAs in ways contributory to intersocietal and intercultural integration. I shall now discuss the channels through which such influences are exerted and give illustrations from other international associations.

The simplest and perhaps most significant proof that ISPAs make a contribution to intersocietal and intercultural integration is the set of consequences among their own members. The very existence of a global association strengthens existing ties across national boundaries and inaugurates new ones. Every ISPA holds world congresses every two, three, or four years. Scientists and professionals who have been accustomed to national meetings find themselves talking shop with their fellows from other nations, making friends, becoming stimulated by ideas and practices in their field stemming from different cultural backgrounds. The contacts initiated at a congress are frequently converted into lasting relationships by common membership in working groups formed to carry on research or policy discussion until the next congress. For most ISPAs it is not the subjects discussed that are significant for global integration, but rather the very fact that persons of high intellectual caliber and prestige are becoming part of the web of intersocietal and intercultural contact under conditions that make for understanding and cooperation. Also, by adding a new level of organization to science and the professions, ISPAs are able to suggest the formation of national associations where none exist, keep national associations abreast of new developments through journals and newsletters, and foster greater continuity of transnational interaction as new recruits come into professions and disciplines. These are the functions that all ISPAs perform for their members. Though they have been illustrated by the two case histories, I shall mention two more associations as further examples.

The effect of the founding of the International Sociological Association (ISA) in 1950 is the subject of a paper by William M. Evan (1975). In it, he explores the hypothesis that "the founding and functioning of the ISA has had a significant impact on the trend toward the internationalization of sociology" (Evan, 1975: 385). By internationalization he means a merging of national sociological communities into an international one in which there is increasing collaboration across borders and acceleration in the growth of sociological knowledge. "Cumulatively, these interrelationships generate feedback effects which strengthen the international association" (Evan, 1975: 386). Evan sets forth a model suggesting that the ISA gives rise to a formal and informal network of communication that reinforces professional norms and values; these, in turn, define problems and formulate and reformulate theory and apply it, resulting in professional collaboration and, finally, in acceleration of the growth of knowledge and its application. Evan is unable to test the hypothesis rigorously, but he performs a content analysis on the papers presented at ISA Congresses from 1953 to 1966 to determine whether they showed an increase in "internationalization." He operationalizes internationalization by combining three indicators into a single index: (1) the degree of foreignness to the author(s) of the languages and publications cited in each article, (2) the degree to which the article is comparative across societies, and (3) the degree of pluralism in the nationality of the author(s) and the locus or loci of the subject matter. This elaborate scheme shows a steady rise in the combined index from 422 in 1953 to 873 in 1966, despite fluctuations in individual series. Evan also regards the increase of ISA research committees from 3 in 1959 to 32 in 1974 as indicative of growing transnational participation.

An ISPA in a very different field, the International Union of Crystallography, has shown similar marked success (Ewald, 1962). Sir Lawrence Bragg, Chairman of the British X-Ray Analysis Group, arranged an international meeting of crystallographers in London in 1946. Those present decided to form the new Union, and the first triennial meeting was held at Harvard in 1948. UNESCO granted a subvention. The journal *Acta Crystallographica* was started and grew rapidly in size and subscribers. The Union has been very active in encouraging improvements in apparatus, the use of computers in research, and the assembly of a world directory of crystallographers. World conferences have grown apace. As with most scientific associations, the main impact has been within the specialized scientific community. There is no doubt that the network of crystallographers has been extended and strengthened—a small but significant contribution to international integration.

A different contribution of ISPAs is that they give wider effect to ethical codes than do national associations. The principal elements of the scientific code enjoin not only scrupulous honesty in reporting data but a search for negative evidence; availability of research results to one's peers freely and promptly; and acknowledgment of indebtedness to others for data or ideas used. These principles are well known among scientists in developed countries, but ISPAs bring them to the attention of scientists in developing countries, especially through world journals. In the case of professional associations, there are often differences in ethical standards among cultures. The international professional association offers at least the opportunity for discussion of a worldwide code. Both scientific and professional associations thus tend to strengthen what little there is in the way of a world normative system.

The World Medical Association affords an example of a professional association that has had wide ethical impact (Bauer, 1957). Its code of ethics, adopted in 1949, has been adopted almost unanimously by national medical associations around the world.

When members of national associations have internalized a code and are seeking to give it effect through governmental programs, their clout is enhanced by their affiliation with the ISPA, especially if the ISPA is carrying on a worldwide campaign. Success in one country can be used as an argument in others. This advantage is more likely to be utilized by professional than by scientific associations, because the former deal directly with the public, are licensed by governmental authorities, and are frequently interested in legislation that will improve the effectiveness of their services. International scientific associations less often feel their objectives need the support of government.

This contribution appeared in the history of the International Council of Nurses. It is also well illustrated by the International Dental Federation (Ennis, 1967). This association had its first meeting in 1901 as a private body very much like the International Statistical Institute. Although it did not become for many years a true federation of national societies, the self-selected Council that controlled it did in fact promote humane policies through the commissions it appointed. Among the successes of the Federation can be noted:

- Encouraging national associations to carry on active campaigns for oral hygiene among school children. So eager was the commission set up by the Federation to do this job that the Council had to restrain it from raising money from the manufacturers of toothbrushes.
- Sending organized relief to 403 Russian dentists after World War I in the form of packages of instruments with which to practice.
- Taking an active role in promoting fluoridation of water supplies.
- Being accepted as a model for national dental associations.
- Actively promoting the training of dental hygienists.
- Advocating dental service for armed forces.

Whether the differentiation of ISPAs into more specialized associations produces globally integrative effects is moot. It undoubtedly makes for more intimate contacts than do immense congresses, and those in attendance are therefore more likely to participate actively. In the *History of the International Congresses of Physiological Sciences,* Wallace O. Fenn (1968) points out that since 1889, when the physiologists first met, pathologists, nutritionists, biochemists, biophysicists, and pharmacologists have all split off and founded their own ISPAs. Despite this, the International Union of Physiological Sciences holds a congress every three years, and each of them is bigger than before the differentiation. One can ask, however, whether differentiation may not sometimes narrow the scope of the conferences to the degree that they become more technical, losing sight of human issues that are brought out in meetings with broader agendas.

This Union illustrates another undoubted way in which ISPAs contribute to intersocietal and intercultural integration. It spends much of its income sponsoring lecture tours of physiologists through regions of the world in which knowledge and research are less advanced, thus building transnational good will while furthering the development of physiology.

Other ISPAs use some of their funds for establishing competitive fellowships for study abroad or stipends for their members to do research abroad or engage in teaching exchanges with foreign colleagues. There is solid evidence that such experiences tend to increase international amity, understanding, and cooperation (Angell, 1969).

By the very nature of the global situation, the international scientific associations that can and do study the world as a unified system are in the natural, not the social, sciences. Geophysicists, oceanographers, meteorologists, and ecologists treat the world as a unity in their everyday work. The parts are seen as linked in a complex whole. Though social scientists speak of the world political, economic, or social system, there is so much conflict among nations that few social scientists think of the world as an actually integrated whole. Thus the internationalization in Evan's article discussed above is less than the global perspective that some natural sciences express. Sciences and international scientific associations can hardly employ concepts of unity in their analysis unless unity in fact exists.

I am not implying that there are no unifying forces at work in the social world, nor that social scientists are unaware of them. There is one circumstance, however, that introduces strain and, ironically, it arises within the social science fraternity. There are strong elements of asymmetry and dependency among social scientists as the heritage of colonialism. The report of a 1973 international seminar, Social Science as Transnational System, describes the problem (Alger and Lyons, 1974). Four main points are made. First, since social scientists in developing countries have been trained mainly in Europe and North America, many of them define a competent scholar as one respected in the country of training. Second, the dependency relation continues because of the inequality of resources for further social science development between the developed and developing countries and the allocation of funds from developed countries to carry on research in developing nations on topics of interest to the developed nations. As one participant put it: " 'All one has here is one-way dependence—the importation of theories, methods and philosophies of science and export of hardly anything, or occasionally of primary products, i.e., raw data, as opposed to the high technology exports of theories and methods.' " Third, inequality of research exchange was expressed by one participant from a developing country as " 'they study us but we don't study them.' " Finally, if social scientists in developing countries want research support from abroad they "have to skew their research to problems such as population and pollution, which they do not consider priority problems. . . . Under such circumstances, it is understandable that several participants raised questions about the desirability of international cooperation in the social sciences" because it "would only strengthen existing patterns of dependency." It was agreed that ISPAs in "helping to bring dominant centers into contact and continuing communication . . . also tend to institutionalize the existing pattern, and . . . to perpetuate, or at least make more rigid, the dependency relations with social scientists outside the dominant centers" (Alger and Lyons, 1974). The Latin American Social Science Council was formed to counter this tendency.

It seems probable that strain is particularly acute in the social sciences, where differences in culture affect the appropriateness of particular methods of research. But even in the natural sciences where this is not so, there are probably elements of dependency. This condition will remain serious in many fields unless strenuous efforts are made to develop strong, autonomous science communities in the developing world. In the meantime, the integrative effect of ISPAs will be retarded.

EXAMPLES OF CONTRIBUTIONS TO IGOs BY ISPAs

The histories of both the International Statistical Institute and the International Council of Nurses mention the contributions they have made via consultative relations granted to them by institutions of the UN system. Since these organizations are charged with carrying out special functions in the interest of peace and security, improving their effectiveness can be a contribution to intersocietal and intercultural integration.

Major assistance of ISPAs to IGOs takes at least the following forms: (1) suggestion of competent personnel to fill posts at IGO headquarters, in the field staff, or on temporary projects; (2) advice on policies and plans; (3) attendance at conferences and workshops when invited; and (4) assumption of responsibility for carrying out projects approved, and sometimes funded, by the IGO.

A typical example of this sort of help is the work of the International Bar Association, which has consultative status with ECOSOC (Lund, 1970). Among its services have been the recommendation that an International Legal Aid Association be founded. This became a reality in 1960. The Bar Association has submitted drafts of conventions, such as "The Administration of Estates Abroad," and has published statements of principle, such as "Protection of Investments Abroad." In 1970, the Association was carrying on studies of international powers of attorney, monopolies and restrictive trade practices, practice of the law by nonlawyers, and professional ethics and professional privilege.

A three-year review by the World Health Organization of the 109 INGOs having consultative relations with it during 1972-1974 affords many other examples. Most of these INGOs are ISPAs but by no means all, e.g., the International Planned Parenthood Federation and the International Committee of the Red Cross. I have chosen the following examples to show the variety of contributions ISPAs are making.[1]

The International Association for Accident and Traffic Medicine arranged joint seminars with WHO on medical examinations of drivers and jointly sponsored with WHO the first International Congress on Vision and Road Safety arranged by *La Prévention Routière Internationale* in Paris.

The International Council on Alcohol and Addictions was represented on WHO Expert Advisory Panels and committees. Members also participated in five working groups. WHO staff members, in turn, have attended the three International Institutes for the Prevention and Treatment of Alcoholism, the second International Conference on Alcoholism and Drug Dependence, the First International Conference on Drug Education, and two International Institutes for the Prevention and Treatment of Drug Dependence.

The International Dental Federation has had a long history of close relations with WHO. Of its former officers, 44 have been appointed to the Expert Advisory Board on Dental Health. WHO and IDF share a common panel of technical experts. Many of them are now engaged in the WHO-IDF International Collaboration Study on Dental Manpower Systems in relation to oral health.

Two ISPAs–the International Federation of Gynecology and Obstetrics and the International Confederation of Midwives–and the Planned Parenthood Federation joined with WHO in a study of the training and practice of midwives and maternity nurses for family planning activities.

The U.S. National Committee of the International Council of Pathology was instrumental in obtaining a grant from the (U.S.) National Cancer Institute to accelerate WHO's program in histological classification of tumors. The grant topped $200,000 by the end of 1974.

Both the International Commission on Radiation Units and Measurements and the International Commission of Radiological Protection have advised WHO, the former on the measurement of radiation dosage and its genetic effects, the latter on standards for means of protection.

The Council for International Organizations of Medical Sciences has closely collaborated with WHO in the preparation of a revision of the International Classification of Diseases. It has also met with WHO personnel on the protection of human rights in the light of scientific and technical progress in biology and medicine.

The International Organization for Standardization (IOS) has worked with the *Bulletin of WHO* in its series of articles on "Programs in Standardization," dealing with standards in public health. IOS has offered to publish, as international standards, any standards established by WHO.

The International Union Against Tuberculosis cooperates in the operation of the International Surveillance Centre, jointly sponsored by the Union, the Organization for Health Research in the Netherlands, and WHO.

These are a few examples of constructive use of consultative relations. During the years 1970-1976, however, there has been much criticism of such relations with ECOSOC and UNESCO, from both sides—the INGOs claiming they are not given enough opportunity to influence the programs of the IGOs, and the IGOs feeling that INGOs do not give enough help to their programs. The basic difficulty seems to be that both IGOs and INGOs want to assert their independence at the same time that they see possible advantages from consultative relations. A seminar convened by the Union of International Associations in 1972 in Milan brought into the open many reservations about the relationship (Deville, 1973). The next year, INGOs reacted furiously when UNESCO informed them that, with the entrance of the Peoples Republic of China into the UN, they would have to drop their affiliated associations in Taiwan if they wished to retain consultative status. This politicization of IGO-INGO relations is inimical to global social integration.

One of the greatest contributions made by ISPAs is the possibility they afford of combining at the world level the abilities of different specialists. This is exercised more often by scientific than by professional associations, since global problems are likely to involve the application of several sciences. There is more often a one-to-one correspondence between professions and particular problems.

The scientific association that has the greatest capacity to arrange coordination of research efforts is the International Council of Scientific Unions (ICSU), founded in 1919. It embraces 17 member unions, each of which either includes several scientific associations or is broken down into divisions, sections, commissions, or committees. Hence, ICSU is a tremendously complex scientific association. It has consultative relations with many of the UN institutions (International Council of Scientific Unions, 1975).

One of the early examples of scientific cooperation set in motion by ICSU was the International Polar Year (1932-1933). This research was carried out cooperatively by the International Meteorological Organization (an IGO), the Permanent Council for the Exploration of the Sea (an IGO), and two constituents of ICSU, the International Union of Geodesy and Geophysics, and the International Union of Radio Science. A more recent example of elaborate collaboration is the Inter-Union Commission on Frequency Allocation for Radio Astronomy and Space Science. This body is composed of members of the International Union of Radio Science, the International Astronomical Union, and the Committee on Space Research of ICSU. It is devoted to obtaining more frequencies from the International Telecommunications Union, the body in ultimate authority. Similar interunion or interassociational efforts are carried on by the Global Atmospheric Research Program, the Scientific Committee on the Problems of the Environment, and the Committee on Water Research.

The International Social Science Council (ISSC) aspires to the same role with its constituents as ICSU has with the ISPAs in the natural sciences, but it is much younger and has many fewer constituents. Like ICSU, ISSC has its headquarters in Paris adjacent to UNESCO. It has consultative relations with UNESCO, ECOSOC, and ILO. The Council has established an exterior body, the European Coordination Centre for Research and Documentation in the Social Sciences (approved by the General Conference of UNESCO in 1962). This Centre has promoted comparative research projects in many fields. Many social science institutes, 238 in all, have participated, and eight volumes of research have been published.

One of the virtues of ISPAs is that they can act more rapidly than can intergovernmental organizations. For example, the ICSU Committee on Contamination by Extra-Territorial Exploration was on top of the problem posed by proposed landings on the moon by manned satellites long before any UN specialized agency became concerned.

Not only have ISPAs had significant roles in supporting institutions of the UN system, they have had influence in creating them. This is seen in the complicated negotiations that went on during the gestation of UNESCO. The *Bureau Internationale de l'Education* was founded as an INGO in 1925. It became an IGO cooperating with the League of Nations in 1929. Also associated with the League before World War II was the *Institut Internationale de Cooperation Intellectuelle* (an IGO). When the League ceased operations in 1939, these two organizations continued precarious independent existences. During World War II, there was established in London a Council of Allied Ministers of Education to plan for postwar educational reconstruction. Both the *Institut* with its cultural (and scientific) focus, and the Bureau with its educational predilections voiced their claims for participation in the planning. In addition, INGOs like the Council for Education in World Citizenship, International Summer Schools, and the New Education Fellowship pushed for a new international initiative. The United States wanted an entirely new UN Organization for Educational and Cultural Reconstruction. ICSU advocated the inclusion of science on a par with education and culture. The upshot of these initiatives, consideration by the Council of Allied Ministers of Education, the Dumbarton Oaks Conference on the UN Organization, and the founding conference of the UN in San Francisco was the conference in London in November 1946 establishing UNESCO.

A simpler case is the transition of the European Molecular Biology Organization (set up as an INGO in a 1963 conference). Its principal aim was to obtain a common laboratory. Not until 1964 did 12 European governments ask the Organization to submit the case for such a laboratory. These governments then signed a treaty setting it up. Thus, after six years as an INGO, the organization became an IGO. In the meantime, it subsisted on foundation grants for continuance of its preparations.

Before concluding this exploration of the contribution of ISPAs to intersocietal and intercultural integration, I should express my suspicion that the more expensive the equipment necessary for fruitful work in a scientific or professional field, the less likely that ISPAs will be the effective means for achieving such work. The book put out by the National Science Foundation entitled *Physics 50 Years Later* gives little indication that ISPAs have been a significant influence in the advances of physics since 1923. International cooperation has been fostered mainly at richly financed intergovernmental institutions like the Center for Theoretical Physics at Trieste or through relations between laboratories such as between the *Centre European de Recherche Nucleaire* at Geneva and the East European Laboratory at Serpushkov in the Soviet Union (with 12 nations represented on each side).

CONCLUSION

I have given examples of the influence of ISPAs on their own members, on affiliated national associations and through them on governments, in the formation of global interdisciplinary committees, as components of unions of ISPAs, and on agencies in the UN system. I suspect, though I have no evidence, that they also affect publics through accounts of their activities in the mass media. For instance, the International Geophysical Year, a project sponsored by ICSU's *Comité Spécial de l'Année Géophysique Internationale* and involving 67 countries in 1957-1958 achieved wide coverage in the world press and undoubtedly prepared the way for the one-world image that was later triggered by the photographs of the earth from spaceships.

Though we cannot now measure the amount of influence toward intersocietal and intercultural integration that flows through these channels, the examples given testify to the significance of ISPAs. They show that ISPAs can and do contribute something over and above what informal networks of scientists and professionals, and national associations contribute. They indicate also that worldwide associations in these fields are solidly organized, that their governing councils and executive committees formulate policies of action, and that their staffs, special committees, and dependent agencies give these policies effect in many ways. In sum, it is clear that ISPAs have become working elements in the world institutional structure that is slowly fostering intersocietal and intercultural integration.

NOTE

1. I am indebted to Dr. M. G. Candau, former Director-General of WHO for supplying me with the organization's mimeographed document, "Triennial Review of Nongovernmental Organizations in Official Relations with WHO."

REFERENCES

ALGER, C. and G. M. LYONS (1974) "Social science as a transnational system." Int. Studies Notes 1, 3: 2-13.

ANGELL, R. C. (1973) "National support for world order: a research report." J. of Conflict Resolution 17, 3: 429-454.

--- (1969) Peace On the March: Transnational Participation. New York: Litton.

BAUER, L. H. (1957) "The World Medical Association: some of its problems and experiences." Int. Associations 9, 11: 773-777.

BEIJER, G. (1972) "The activities of the Research Group for European Migration Problems." Int. Associations 17, 9: 509-513.

BRIDGES, D. C. (1967) A History of the International Council of Nurses: The First Sixty-Five Years. Philadelphia: J. B. Lippincott.

DEVILLE, G. (1973) "The philosophy of nongovernmental organization." Int. Associations 25, 3: 147-161.

EVAN, W. M. (1975) "The International Sociological Association and the internationalization of sociology." Int. Social Sci. J. 27, 2: 385-393.

ENNIS, J. (1967) The Story of the Fédération Dentaire Internationale. London: International Dental Federation.

EWALD, P. P. (1962) Fifty Years of X-Ray Diffraction. Utrecht: N.V.A. Oothoek's Uit Gevermatatschappij.

FENN, W. O. (ed.) (1968) History of the International Congresses of Physiological Sciences. Baltimore: American Physiological Society.

International Council of Scientific Unions (1975) Yearbook 1975. Paris: ICSU Secretariat.

KRIESBERG, L. (1972) "International organizations and transnational integration." Int. Associations 24, 11: 520-525.

LUND, T., Sir (1970) The Association: Its Members and Their Activities. London: International Bar Association.

Nature (1969) "EMBO: an international laboratory needed." 222, 5: 836.

NIXON, J. W. (1966) A History of the International Statistical Institute. The Hague: International Statistical Institute.

12

INTERNATIONAL ORGANIZATIONS OF TEACHERS: Their Policy Impact

William G. Carr

The impact of organized teachers on United Nations agencies and other intergovernmental organizations is very recent. This chapter provides a brief history of teachers' organizations and their international impact. The discussion will consider: (a) early efforts of these organizations, (b) the creation of the UN and UNESCO, (c) current international teachers' organizations, (d) their impact on the UN system, and (e) several tentative conclusions.

EARLY EFFORTS BY TEACHERS' ORGANIZATION TO INFLUENCE INTERNATIONAL AGENCIES

THE INTERWAR PERIOD

The International Bureau of Education (IBE) was founded at Geneva in 1925 to exchange information about elementary and secondary schools. Governments did not become "members" of IBE until 1929. As late as 1946 the Bureau included as members only fifteen governments—and these fifteen did not include either the United States or the United Kingdom. However, about fifty governments, including that of the United States, sent delegations to the annual International Conferences on Instruction which the IBE sponsored. The IBE, with its International Conferences, has now become an agency of UNESCO, retaining, however, a considerable measure of autonomy.

Governments, not teachers' organizations, established and operated the IBE. Occasionally, officers of national teachers' organizations have been included in some national delegations to the IBE Conferences. International teachers' organizations have been routinely invited to send official observers to the IBE Conferences and vice versa. These invitations are often accepted but the contacts between IBE and teachers' organizations, national or international, have been marked as a rule by courteous and benevolent interest.

Efforts by teachers' organizations to secure a reference to education in the text of the League of Nations Covenant were persistent, sincere, and ineffective. At Paris in 1919, Dr. Fanny Fern Andrews, a spokeswoman for the National Education Association of the United States and The National Union of Teachers of England and Wales, was accorded a brief hearing on this subject in a parlor of the Hotel Crillon, where Woodrow Wilson and his colleagues from France, Great Britain, and Italy were completing their work on the text of the League Covenant. Although her case for including education in the scope of the League's interests was clearly worded, eloquently expressed, and politely heard, it was nevertheless rejected. In 1921, the League of Nations Assembly deleted the word "education" from one of its documents with only one vote (that of Haiti) opposing the action (Carr, 1945; Council for Education in World Citizenship, 1943; Rosello, 1943).

WORLD WAR II

During the last years of World War II, an increasing concern was developing among teachers in the allied countries that the power of their profession should be enlisted to help deal constructively with international problems. Informal discussions began during 1942 on a variety of problems relating to organized education. How could the misuse and subversion of education as an instrument of deceit and racism be prevented? How could educational services in the heavily damaged war areas be restored? Universities had been destroyed, the preparation of teachers neglected, libraries burned, school buildings demolished, the schooling of an entire generation interrupted and distorted. How could education be organized for the tasks of reconstruction and international cooperation?

In 1943, two small books on postwar education were published almost simultaneously, one in the United States and one in Great Britain (Council for Education in World Citizenship, 1943; Educational Policies Commission, 1943). The American report was issued by the Educational

Policies Commission; the British report was sponsored by the Council on Citizenship. Although they were independently prepared, the two documents have much in common. Both were prepared by nongovernmental agencies and, with minor variations, their recommendations are much alike. Both considered the reconstruction of schools in the war-devastated countries, the control of education in the Axis countries after the war, and the establishment of international organizations for international cooperation in education to help keep the peace.

The American report, *Education and the People's Peace,* became a primary topic of conversation among teachers, school administrators, and the general public. A substantial campaign, financed by the War and Peace Fund of the National Education Association, was launched to secure public consideration of the recommendations.

The interest in this topic among civic and scholastic groups became intense as additional segments of the public became involved. The American Academy of Political and Social Science devoted its *Annals* for September 1944 to the topic, with a score of contributed articles on various aspects of the (then) novel idea of organized education as an instrument for international understanding and peace (Carr, 1944).

So many different professional groups became interested that early in 1943, a Liaison Committee on International Education was formed.[1] Among its other activities, the Liaison Committee and its member associations organized a series of three International Education Assemblies.

The first such Assembly, meeting at Harper's Ferry, West Virginia in September 1943, brought together participants from 26 countries (International Education Assembly, 1943, 1945a). These were in no sense delegates of governments, but were men and women who happened to be in the United States and who held positions of distinction and leadership in the educational circles of their respective countries. The Harper's Ferry Assembly agreed on the need for two new international agencies: (1) a permanent postwar international organization to encourage peace through educational and cultural exchanges; and (2) a temporary postwar agency to deal with such immediate issues as the ending of the fascist school systems and the reconstruction of schools and colleges destroyed during the war.

The Second Assembly met at Hood College, Frederick, Maryland in June 1944 (International Education Assembly, 1944). The central task of this Assembly was to formulate a statement describing "Education in a Free Society." The participants from thirty countries succeeded in their ambitious, self-imposed task, the first such effort on an international basis in history. The Assembly was in session on D-Day, the invasion of German-occupied Western Europe. The hopes and anxieties evoked by this event led the Assembly to renew the proposal for prompt establishment of an international agency to promote educational relations throughout the world.

The third and final session of the International Education Assembly was held in New York City in April 1945 (International Education Assembly, 1945b). By that time, the number of countries from which participants were drawn had risen to 34. At the time the Assembly convened, it was known that an international meeting of government-appointed delegates would in a few weeks begin its work on a United Nations Charter. It was also known that the "Dumbarton Oaks Draft" which would be the principal working paper at San Francisco had completely avoided the question of international cooperation in education.

As the news of D-Day had solidified the cooperative spirit of the Assembly in 1944, so the shattering news of the sudden death of President Roosevelt gave a sense of increased unity of purpose to the Assembly in April 1945. A strong resolution was framed, adopted, and sent to delegates who would soon assemble in San Francisco. It called for an International Office of Education to be a clearinghouse and advisory center for educational activities, especially relevant to fostering international understanding, goodwill, and peace.

Thus, the impact of organized educators on the San Francisco Conference was transmitted primarily through the Liaison Committee for International Education and the International Education Assembly. The National Education Association (1945) generated the basic concepts through its Educational Policies Commission. The NEA also contributed the principal funds and staff through the NEA War and Peace Fund. The entire effort was implemented through United States organizations. There is no record of comparable or parallel activity on the part of teachers' organizations in any other country. Even the International Education Assembly, although it included nationals of many other countries, was organized, led, and financed by United States personnel and organizations.

THE CREATION OF THE UN AND UNESCO

THE UN CHARTER

An important new feature marked the composition of the U.S. delegation at San Francisco. In addition to the traditional array of delegates, alternates, and experts, a new post called "consultant" was created. Private U.S. organizations, 42 in all, were selected by the State Department and invited to name a consultant to the U.S. delegation. This arrangement is, so far as I know, unique. It has not been tried by other governments or on any occasion except the San Francisco meeting. The motive of the Department was to help insure Senate ratification of the UN and to secure suggestions for the text of the Charter. The influence of the consultants was most notable in the promotion of educational cooperation (UN Charter, article 55), the definition of human rights (article 68), and the formal recognition of nongovernmental organizations (article 71).

The outlook for securing recognition for education in the text of the United Nations Charter was not favorable. The delegations of such great powers as the Soviet Union, the United Kingdom, the United States, and France showed at the outset little or no interest in the subject. The delegation of China, however, on the opening day of the San Francisco Conference issued a formal proposal that the UN Charter should "specifically provide for the promotion of educational and other forms of cultural cooperation" (Eisenhower, 1962).

The first suggestion of educational cooperation in a plenary session of the Conference occurred on April 28 when the delegate from Lebanon said, "Educational and intellectual cooperation among the nations is of primary importance." Later the same day, the spokesman for Uruguay called for the establishment of an international council on education. Similar points were made during the opening general discussions by delegations from Ecuador, Greece, Norway, Panama, Philippines, Iran, and Haiti.

Thus, in the opening "general debate" only the small nations expressed interest in the educational and cultural functions of the new organization that was being built. Possibly these delegations knew that they could not function with decisive power in military and economic matters but hoped that in educational and cultural matters they could more nearly approximate the influence of the greater powers.

In April 1945, the National Education Association conducted a public opinion poll on international cooperation, using some of the resources of its War and Peace Fund. The poll suggested that about 84% of the American people responded affirmatively to the question of establishing a world agency for education. On May 1, at a meeting of the United States delegation, this poll and its results were called to the attention of the delegation.

The difficulties in securing a reference to education in the UN Charter continued. They centered around these four points:

(1) The fear of the leaders of the U.S. delegation that the U.S. Senate might withhold its consent from the Charter as it had done in 1919 when offered the League of Nations Covenant. This apprehension was genuine and widespread. Charter reference to education, it was feared, would be a liability in seeking the advice and consent of the Senate.

(2) The argument of John Foster Dulles, at that time an influential and respected U.S. delegate at San Francisco, that the word "cultural" was less likely to create objections than the word "educational." At the same time, he held that "cultural" adequately covered the kind of activity the teachers had in mind (or, at least, what they ought to have in mind).

(3) The pressure of time which made it difficult to secure attention to any topic that was not already embodied in the Dumbarton Oaks draft document.

(4) The doubt that the general public really cared much, one way or the other, about international cooperation in education.

The NEA opinion poll, mentioned above, was helpful in offsetting the effect of the alleged public indifference. Furthermore, a variety of arrangements were made in San Francisco to ensure that the U.S. delegation heard from the education community and from others interested in education. There is no way to measure the success of these efforts. They were certainly extensive and they certainly represented the payoff on the vigorous public information campaign that the NEA had conducted over the preceding three years (World Confederation of Organizations of the Teaching Profession, 1961).

On May 4, 1945, just under the agreed time limit, the joint proposals of amendments to the Dumbarton Oaks document from the United States, China, Great Britain, and the Soviet Union were submitted. The joint proposals for amendments lacked any reference to education. This was a most discouraging event. It was not, however, the last word on the subject; it was a reversal but not a catastrophe.

In Washington, the NEA arranged for Senators J. William Fulbright of Arkansas and Robert A. Taft of Ohio to sponsor a resolution in the Senate favoring the participation of the United States in an international office of education and cultural development. Sponsorship on a bipartisan basis by two highly influential Senators was a major advantage. In the House of Representatives Congressman (later Senator) Karl Mundt introduced a similar resolution (United States Congress, 1945). A few weeks earlier, Congressman Mundt, a former public school administrator, had delivered the Keynote speech to the International Education Assembly meeting in New York.

In San Francisco, meanwhile, the daily meetings of the consultants with the policy-making members of the U.S. delegation continued. On the question of educational cooperation, a group of consultants was formed representing Agriculture, Business, Labor, and Education. This group was usually referred to by the acronym ABLE. The representative of the National Council of Churches said later that, if Religion had been included, the ABLE group would have been ABLER; he was probably correct.

On the evening of May 12, 1945, during a panel discussion on the NBC network, the question of international cooperation in education was brought to a considerable segment of public opinion (University of the Air of the National Broadcasting Company, 1945). During this broadcast, Assistant Secretary of State Archibald MacLeish asked me to "state the case" for educational cooperation. I began by asking another panel member, Delegate Virginia Gildersleeve, why the United States had failed to support the Chinese proposal that the UN sponsor "educational and other forms of cultural cooperation" and why the U.S. delegation had deleted the specific reference to education. She replied that "education" for some people might have connotations of propaganda.

> *Carr:* Well, of course, some dictionary definitions for education do mention propaganda. But some dictionary definitions of culture mention bacteria. After all, a recent public opinion poll showed that 84% of the American people wanted an international agency *for education.* Is there any other issue on which you could get such nearly universal approval?
>
> *Dean Gildersleeve:* I don't know of any, Dr. Carr.
>
> *Carr:* What then accounts for the fear of the word "education"?
>
> *Dean Gildersleeve:* I suppose it springs from a variety of things . . . possible interference with our American system of education by foreign countries . . . I don't share this viewpoint, but apparently it does exist. We might ask Mr. MacLeish to define "culture."

Mr. MacLeish caught the lateral pass and ran with it. "Culture," said he, "is one of the things you can't define . . . you describe it." He then described it eloquently but the exercise produced no definition. It then produced the following exchange:

> *Carr:* Mr. MacLeish, isn't "culture" a rather cloudy term to the man on the street? Might it not also mean different things to different nations? If we mean to include education let us say so, as the Chinese proposed, by explicit reference. . . . Mutual understanding, based on educational cooperation, is vital to the success of the world organization that the Conference is building. Educational cooperation is not a minor gadget to be added to the machinery of international life later on. It is the motor which makes the machine run.
>
> *Secretary MacLeish:* You know, that's converted me. . . . Without that we'll never obtain a lasting peace.

On May 16, the ABLE group met at length with the entire U.S. delegation to discuss our collective views on education's role in the Charter. On the same day the House of Representatives Committee on Foreign Affairs approved Congressman Mundt's resolution.

In fact, May 16 was the day the breakthrough became visible. The San Francisco *Chronicle* declared that the role of education had become a crucial issue at the Conference. The San Francisco *News* predicted (correctly) that a United Nations education conference would be held within the next few months. The New York *Times* said that discussion about education was one of the most significant developments of the Conference (United Nations Conference on International Organization, 1945).

The Mundt resolution passed the House of Representatives without a dissenting vote. The Taft-Fulbright resolution passed the Senate unanimously. On May 22, a definite proposal for the promotion of educational cooperation (the *promotion,* not merely the *provision*), was offered to the Drafting Committee by the United States delegation. It was unanimously adopted.

Another education facet in the United Nations Charter should be mentioned in passing. The "Big Four" joint proposals included a new and additional section in the Charter to create a Trusteeship Council to deal with those parts of the world which were then in colonial status. The draft, as submitted, contained no reference to the necessity of education as a means to prepare the people of colonial areas for either independence or self-government. It was pointed out that the best test of the good faith of a trustee nation would be its policy regarding the education of the inhabitants of the trust territory. A trustee could provide food, shelter, and health services without preparing the recipients for self-government. The proposition was so evident that a revision of the draft Charter was easily secured.

Thus, the United Nations Charter became the first great international document to give explicit recognition to the power of education to help keep the peace. Without the persistent efforts of the professional organizations to which I have referred, such provisions would not occur at nine places in the Charter.

THE UNESCO CONSTITUTION

While the UN Charter-writing activities were going forward in Washington and San Francisco, related activities were underway in London. There, in cooperation with the United Kingdom Ministry of Education, a Council of Allied Ministers of Education had been established. This Council consisted mainly of Ministers of Education attached to several European governments-in-exile which were sheltered in London when their countries were occupied during World War II. San Francisco having established the general principle that the United Nations system would concern itself with educational cooperation, it was necessary and natural to design an organization to conduct these future intergovernmental activities in education.

The UN Conference for the Establishment of an Educational and Cultural Organization opened in London on November 1, 1945 (UNESCO, 1946). The National Education Association was represented by its president, who served as a member of the U.S. delegation. I attended the London Conference as part of the Secretariat, having been designated by the U.S. State Department, at the request of the British Government, to serve as deputy to Sir Alfred Zimmern, the Secretary-General of the Conference. Excepting the National Education Association of the United States, no teachers' organization was represented in any other delegation at the London Conference.

Among other duties assigned to me by Sir Alfred, I was asked to staff the Committee which drafted the Preamble and the statement of purposes of the Educational and Cultural Organization. In this capacity, I tried to make sure that the central purpose of UNESCO would be consistent with the objectives foreseen by the Educational Policies Commission in 1943. The Constitution of UNESCO as finally adopted contains in Article I a clear and definite statement of its *purpose*: "to contribute to peace and security." The same Article specifies that the *method* by which UNESCO will seek to implement its purpose is international collaboration through education, science, and culture. The word "purpose" is singular. No other purpose is named in any article of the Constitution. The purpose was chosen after extensive discussion in the Drafting Committee, and then approved by vote of the First Commission, the General Drafting Committee, and finally by the Conference as a whole.

So far we have considered the impact of organized teachers on (1) the establishment of the United Nations, (2) the inclusion of education in the UN framework, and (3) the establishment of UNESCO. In these policy decisions, the influence of organized teachers in the United States was sustained chiefly, but not exclusively, through the National Education Association. They formulated the basic policy through the Educational Policies Commission, provided the necessary resources for advancing these policies to the point of action through the NEA War and Peace Fund, and took the lead in organizing and maintaining the International Education Assembly.

CURRENT INTERNATIONAL TEACHERS' ORGANIZATIONS

We now turn attention to the formation of international teachers' organizations and the impact of these international agencies on the work of the United Nations, UNESCO, and to some extent on other agencies within the United Nations system. It will not be possible in this chapter to recount the history of these international agencies, nor is it necessary to do so. It will suffice to discuss the three largest and most influential of the approximately dozen international teachers' organizations.

1. The International Federation of Teachers Syndicates (usually known as FISE, its acronym in the French language) consists chiefly of the national syndicates of employees of the ministries of education in the communist countries, the USSR, Poland, East Germany, Czechoslovakia, Rumania, Bulgaria, and Cuba. The teachers of the Peoples Republic of China used to be members of FISE but, since the Sino-Soviet political schism, the teachers of mainland China have not been identified with FISE or with any other international organization. The syndicate of teachers of Mexico has been an active member of FISE since 1946. The teachers' syndicates of most of the Arab states have been intermittently affiliated to FISE.

2. The International Federation of Free Teachers' Unions IFFTU) is composed of some, but not all, of the teachers' unions affiliated with the general labor movements of their respective countries. The larger members of IFFTU include the American Federation of Teachers (AFL-CIO), and the teachers' unions of Belgium, Germany, and a few other countries.

3. The World Confederation of Organizations of the Teaching Profession (WCOTP) consists of about 125 teachers' organizations in about ninety countries, widely distributed throughout the world. Some of its members are affiliated with trade unions, some are independent. Some of the member organizations, especially in Europe, are limited to teachers in the elementary schools, some to teachers in secondary schools, and some are all-inclusive. Its largest member organizations are the National Education Association of the United States, the Japan Teachers Union, the All-India Federation of Education Associations, the syndicate of elementary school teachers of France, the Teachers' Organization of the Republic of Indonesia, the Canadian Teachers' Federation, the National Union of Teachers of England and Wales, the Nigeria Union of Teachers, the Philippine Public School Teachers' Association, the West German Teachers' Society, and the Australian Teachers' Federation. In addition to these very large organizations, WCOTP has in its membership many small national organizations such as those of Malta (280 members), Botswana (265), Barbados (222), Tonga (222), Belize (200), Dahomey (200), Iceland (200), Singapore (150), Guyana (120), and the British Virgin Islands (80) (World Confederation, 1952-75, 1947).

Among these international bodies, the WCOTP most completely and consistently shows a continued and effective interest in, and influence on, the United Nations and its specialized agencies. WCOTP is the only teachers' organization to maintain a full-time regional office in Paris for day-to-day cooperation with UNESCO and its Secretariat.

In the years immediately after the war, WCOTP was actively cooperating in the UNESCO program of relief and rehabilitation. It encouraged its members to buy UNESCO coupons which helped to restore damaged or destroyed textbooks, libraries, and teaching materials. It worked closely to support the United Nations Children's Emergency Fund (UNICEF).

These three are currently the major teachers' organizations at the international level. Others are more specialized or regional, e.g., the World Union of Catholic Teachers, the (Latin) American Teachers' Congress, a newly established African continental teachers' organization, the Federation of Arab Teachers' Syndicates, and the Caribbean Union of Teachers. There is also an International Association of University Professors and Lecturers.

IMPACT ON THE UN SYSTEM

The influence of WCOTP and other international organizations on the UN system includes the following areas, which will be discussed in the balance of this chapter:

(1) The Universal Declaration of Human Rights
(2) The International University
(3) The International Recommendation on the Status of Teachers

THE UNIVERSAL DECLARATION OF HUMAN RIGHTS

Between 1946 and 1948, the World Confederation urged that education be one of the Human Rights in the Universal Declaration. The success of the effort was due in part to the fact that the drafters of the Universal Declaration did not resist the proposal. However, after its adoption, the dissemination and implementation of the Declaration was diligently and quite effectively promoted by international teachers' organizations. Consideration of the Universal Declaration as an aspect of general education has been substantially extended as a result of the curriculum-making aspects of the work of teachers' organizations.

The construction of the Berlin Wall in August 1962, ending the massive westward exodus of teachers and other trained workers from East Germany, gave WCOTP an opportunity to deal with Human Rights in more specific terms. In 1959, a special WCOTP committee delivered a report on refugee teachers from the East German government (World Confederation, 1959). Between 1947 and 1948, the report showed, the number of refugee teachers from East Germany increased by 50% in that single year. The migration and the Wall were both suggestive of a violation of the UN Declaration of Human Rights. The 1959 report of the Special WCOTP

Committee on Refugee Teachers also dealt with the oppression of dissident students; including adverse discrimination because of the occupation or former occupation of their parents or their lack of political orthodoxy. Teachers particularly resented the requirement that they present false or misleading materials to their students, spy and report on pupils and their parents, attend numerous political rallies, and recruit students for the Young Pioneers and other communist youth programs at the expense of activities sponsored by the church and approved by the home. The 1959 Delegate Assembly of WCOTP adopted this report and sent it to the United Nations, to UNESCO, and to the media for maximum worldwide distribution.

The movement of teachers, physicians, engineers, technicians, and others out of East Germany continued. It was found necessary to establish three large refugee centers in West Berlin. The number of teachers passing through these centers averaged 100 a month for seven successive years.

Unwilling or unable to cease the measures which were driving intellectual leaders out of their country, the East German government reacted savagely. On August 13, 1961, a ten-foot wall topped with barbed wire was constructed along the westward edge of the East Berlin zone. The East German government said at first that the purpose of the Wall was to exclude "spies and saboteurs" from the West. However, the reason given was so unbelievable that in a few months it felt obliged to admit that the chief reason for the Wall was to restrain potential defectors from East Germany.

The German teachers, supported by other WCOPT members in Western Europe, requested that WCOTP establish a special committee to study this development on the spot and to issue a report. The WCOTP President, Sir Ronald Gould, appointed me to chair this Committee, which also included representatives of WCOTP member organizations in Kenya, Malaysia, and France. The Committee, meeting in Berlin for a week beginning on December 6, 1961, took evidence from about 100 witnesses and examined about 200 teachers, retired teachers, and students from East Berlin (World Confederation, 1961). The Committee's first action was to invite the East German Teachers' Union to supply information in person or in writing. This invitation, although delivered and received, remained unanswered. The 300 pages of testimony collected were filled with hope and despair, courage and inspiration. The Wall had separated teachers from students and students from their families. The situation for many teachers obliged to remain in East Germany constituted a grave assault on their professional and human rights (World Confederation, 1961).

The Committee cited the Universal Declaration of Human Rights issued by the United Nations on December 10, 1948, and in particular Article 13 of that document which states that: "Everyone has the right to leave any country, including his own, and to return to his country."

The Committee recommended that its report be given the widest possible distribution so that the catastrophic events it reports "may reach as many understanding minds and compassionate hearts as possible." This recommendation was carried out in a variety of ways. For example, WCOTP arranged for three East German teachers to tour Latin America and West Africa to give a first-hand account to teachers and their organizations, as well as to the general public in those regions.

THE INTERNATIONAL UNIVERSITY

Speaking in Stockholm to the 1961 Delegate Assembly of WCOTP, President Dwight Eisenhower suggested that teachers develop, with representatives of the United Nations, plans for an international university with a faculty of outstanding scholars selected by the UN General Assembly from lists nominated by the international teachers' organizations. Suggestions of this kind had appeared before in the writings of educators, scholars, and political leaders. In 1919, a short-lived International University was established in Brussels but it was discontinued because of a lack of funds in the mid-1930s. Bertrand Russell in 1942, Sir Alfred Zimmern, Albert Einstein in 1949, are a few of the well-known people who have voiced an interest in this topic (Zweig, 1967).

The WCOTP Executive Committee moved at once to pursue the idea suggested to the Confederation by President Eisenhower. Contacts were made at once with the United Nations Secretariat. The Executive Committee expressed its endorsement of the basic idea, offered to cooperate in every possible way, and proposed tentatively that the new institution might well be located in Geneva and be limited at first to a one-year postgraduate course for not more than 500 students. Since then, the idea has moved slowly forward toward a visible and tangible institution. No doubt the forum thus provided for President Eisenhower on this theme and the subsequent worldwide publicity concerning his observations made a substantial contribution to the advancement of the idea, and the eventual approval in 1973 of a United Nations University by the General Assembly.

THE INTERNATIONAL RECOMMENDATION ON THE STATUS OF TEACHERS

The most visible single result of the impact of teachers' organizations on agencies within the UN system, is the 1966 UNESCO Recommendation on the Status of Teachers. The Recommendation deals with teachers in both public and private schools up through the completion of secondary education, including technical, vocational, and art education. It includes statements in educational policies; preparation for the profession (selection and teacher preparation programs and institutions); further education for teachers; employment and careers (entry into the profession, advancement, promotion, tenure, disciplinary procedures related to breaches of professional conduct, health examinations, women teachers with family responsibilities, and part-time services); rights and responsibilities (professional freedom and responsibility); conditions for effective teaching and learning (class size, ancillary staff, teaching aids, hours of work, holidays, study and special leave, sick leave, maternity leave, teacher exchange, special provisions for teaching in remote and rural areas); teachers salaries; social security (medical care, employment injury benefits, old age benefits, invalidity benefits, survivor's benefits); teacher shortage and supply (UNESCO, 1966).

The achievement of this Recommendation involved activities by international teachers' organizations which extended over a period of about eleven years.

The theme of the Istanbul Assembly of WCOTP in 1955 was "The Status of Teachers." The membership of WCOTP at that time was beginning a rapid expansion, particularly in Asia and Africa. It was decided to move towards the achievement of global acceptance of a Charter of the rights and responsibilities of teachers. In view of the growing membership it appeared that the global objective should be first approached through a series of regional surveys and conferences.

In 1959, following several exploratory meetings, WCOTP established a Commission on Educational Policy for Africa. Next year, meeting in Uganda, this Commission concluded that the status of the teaching profession should be known not only by teachers themselves, but also by governments and by the public at large.

Regional meetings for Asia and the Americas followed. These sessions were similar in purpose to the African meetings but were adapted to the conditions and needs in the other regions. The Asian organizations met in Malaya; the Americans met in Costa Rica. These meetings also concluded that the effective use of education as a means of national development required public recognition of the professional status of teaching.

These early efforts in regional planning met some difficulties, chiefly because the facts concerning the status of teachers had not been assembled, arranged, and distributed. Accordingly, in 1961, WCOTP initiated a series of regional surveys of the status of teachers, together with regional conferences to disseminate the findings and consider their implications (World Confederation, 1952-1975). The first of these surveys was carried out by WCOTP in Africa in 1961, with some financial assistance from UNESCO and the Ford Foundation. A former president of the

Gambia Teachers Union, after bilingual studies and visits in 26 African countries, compiled the pioneer survey, *The Status of the Teaching Profession in Africa.*

The next step in Africa was taken in 1963 when WCOTP organized a conference in Niger. Here, for the first time in African history, government representatives and leaders of the teachers' organizations from the same countries met to consider jointly the teaching profession in Africa. The results of their conference were published by the World Confederation (1952-1975) in a substantial *Handbook for Raising Teacher Status in Africa.*

In Asia, the survey of the Status of the Teaching Profession was conducted by the Director of the WCOTP regional office in New Delhi. It covered fourteen Asian and Pacific countries including Austalia and New Zealand. As in Africa, the investigation was assisted by a grant from UNESCO. An Asian regional expert committee reviewed the study and prepared recommendations. The study, including the recommendations, was circulated to all Asian ministries of education as well as to the UN and UNESCO.

The *Status of Teachers in the Americas, from Argentina to Canada,* was published in 1964 and was followed in 1966 by a report on the status of teachers in seventeen small countries and islands in the Caribbean. Special studies were also made on primary and secondary teachers and on teachers of health and physical education. As all these studies were completed and distributed, contacts were maintained with the major intergovernmental organizations and especially with UNESCO.

The activity of UNESCO in regard to the status of teachers began in 1947 when its General Conference, meeting in Mexico City, called for the preparation of a Teacher's Charter. The international teachers' organizations were enthusiastic in their support of the teacher's charter idea. However, action on the 1947 directive was repeatedly postponed. The lack of full unity among the teachers' organizations of the world is probably responsible for most of the delay. Besides, the word "Education" in the UNESCO title was not as a rule perceived as having any connection with the grubby work of the ordinary classroom teacher.

In fact, the credit for the first active intergovernmental leadership regarding the status of teachers must be accorded to the International Labor Organization. The ILO in 1958 convened a meeting on "The Economic and Social Conditions of Teachers." This, of course, could and should have been done by UNESCO at any time in the intervening eleven years between the Mexico City directive to UNESCO and the convening of the ILO meeting in Geneva. However, the UNESCO and ILO leadership apparently decided to avoid jurisdictional conflict on the question. Thus, the further expert meetings were jointly sponsored by the agencies and the expert meetings alternated between the two headquarters, Paris and Geneva. The participants in these meetings were nearly all leaders of national and international teachers' organizations.

After all these years of preparation, an Intergovernmental Conference of an International Recommendation Concerning the Status of Teachers was convened in Paris by UNESCO in September-October 1966. It seems clear that this International Recommendation came into existence as the direct result of the efforts of internationally organized teachers.

TENTATIVE CONCLUSIONS

(1) Although many *national* teachers' organizations are well over 100 years old, they have become somewhat influential in the affairs of intergovernmental organizations only in the past thirty years or so. However, a general and sincere desire to promote a peaceful world has characterized some national teachers' organizations for a much longer period than thirty years.

(2) The oldest international teachers' organizations have been in existence for about fifty years. The same general observations apply to them as to the national organizations just mentioned.

(3) The first clear-cut illustration of the influence of teachers' organizations on intergovernmental agencies was the inclusion of explicit references to education in the UN Charter Conference at San Francisco in 1945. This success rested upon about five years of intensive publicity and preparation.

(4) International teachers' organizations today have considerable influence on the program and operations of UNESCO—and to a lesser extent on the work of ILO, FAO, UNICEF, and the United Nations University.

(5) The effectiveness of international organizations is limited by two major circumstances. First, like almost every agency connected with education, teachers' organizations have a chronic lack of funds; second, the shortage of resources is aggravated by the lack of unity. When intergovernmental organizations wish to learn the views of the teaching profession, there is no single source to which they may turn. If, as sometimes happens, they seek the advice of only one of the existing international organizations of teachers, the intergovernmental organizations can count on trouble from the organizations they did not consult.

(6) To influence the policies of intergovernmental organizations, international teachers' organizations need to concert their efforts directly with respect to these agencies, as well as indirectly. In the latter case, national teachers' organizations must attempt to persuade officials of their own governments to exert influence on the policy-making process of intergovernmental agencies.

It is sometimes difficult, but always important, to try to distinguish activity from influence. International nongovernmental associations tend to have more of the former than of the latter. Nevertheless, NGOs have exerted some influence. Their activity is, in most cases, a precondition for the exercise of influence.

NOTE

1. For the record, the organizations whose representatives composed the Liaison Committee for International Education were:

American Association for Adult Education
American Association of Teachers Colleges
American Association of University Women
American Council, Institute of Pacific Relations
American Council of Learned Societies
American Educational Research Association
American Federation of Teachers, Commission on Post-War Planning
American Library Association
Association of American Colleges
Association of Colleges and Secondary Schools for Negroes
Commission on Resources and Education, National Education Association and Progressive Education Association
Institute for the Advancement of Visual Education and Vocational Training
Institute of International Education
International Council on Religious Education
Modern Language Association of America
National Association of Secondary School Principals
National Catholic Education Association
National Council of Teachers of English
National Council for the Social Studies
National Education Association, Commission on the Defense of Democracy through Education, Committee on International Relations, Departments of Adult Education, Elementary School Principals, and Supervision and Curriculum Development
National Society of College Teachers of Education
Progressive Education Association
United States Committee for Intellectual Cooperation
World Federation of Education Associations

REFERENCES

CARR, W. G. (1945) Only By Understanding. Foreign Policy Association Headline Series 52. New York: Foreign Policy Association.

--- [ed.] (1944) International Frontiers in Education. Annals of the Academy of Political and Social Science (special issue) September.

Council for Education in World Citizenship (1943) Education and the United Nations. London: Author. (Republished in the United States by the American Council on Public Affairs, Washington, DC.)

Educational Policies Commission (1943) Education and the People's Peace. Washington: National Education Association.

EISENHOWER, D. D. (1962) "Address to the Eleventh Delegate Assembly of WCOTP," in World Confederation of Organizations of the Teaching Profession Annual Report, 1962. Washington, DC: World Confederation.

International Education Assembly (1945a) Education in the United Nations. New York: The School Executive.

--- (1945b) International Education Through Cultural Exchange (New York City Meeting). New York: The School Executive.

--- (1944) Education for a Free Society (Hood College Meeting). Washington, DC: The Assembly.

--- (1943) Education for International Security (Harpers Ferry Meeting). Washington, DC: The Assembly.

National Education Association (1945) The NEA at the United Nations Conference. NEA Leaders' Letter 40. Washington, DC: Author.

ROSELLO, P. (1943) Les Precurseurs du Bureau International d'Education. Geneva: International Bureau of Education.

United Nations Conference on International Organization (1945) Journal 26: Author.

United Nations Educational, Scientific and Cultural Organization [UNESCO] (1966) Recommendation Concerning the Status of Teachers, Special Intergovernmental Conference. Paris: Author.

--- (1946) Conference for the Establishment of the United Nations Educational, Scientific and Cultural Organization, Preparatory Commission. London: Her Majesty's Stationery Office.

United States Congress, 79th Congress (1945) House Joint Resolution 305 and Senate Joint Resolution 135.

University of the Air of the National Broadcasting Company (1945) Report from San Francisco, III, Program 13, May 12. New York: NBC.

World Confederation of Organizations of the Teaching Profession (1952-1975) Annual Reports. Morges, Switzerland: Author.

--- (1963) Handbook for Raising Teacher Status in Africa. Washington, DC: Author.

--- (1961) Inquiry Commission on Berlin. Washington, DC: Author.

--- (1959) A Call to Freedom: Report Submitted to the Assembly of Delegates, July 31-August 7. Special Committee on Refugee Teachers from East Germany. Paris: Author.

World Organization of the Teaching Profession (1947) Proceedings of the World Conference of the Teaching Profession, August 17-30, 1946. Preparatory Commission. Washington, DC: Author.

ZWEIG, M. (1967) The Idea of a World University. Carbondale: Southern Illinois Univ. Press.

13

HUMAN RIGHTS ORGANIZATIONS

Harry M. Scoble and Laurie S. Wiseberg

In attempting to assess the state of human rights—and thus the policy impact of international nongovernmental human rights organizations—in the late 1970s, one faces the Dickensian paradox that these are the best of times; these are the worst of times. That is to say, compared with only thirty years ago, one finds an emergent corpus of international law specifying human rights (Sohn and Buergenthal, 1973; Brownlie, 1971); nascent international governmental institutions devoted to the protection of these rights; and, especially, an elaboration of political interest group organizations dedicated to the effective maintenance and expansion of human rights—leading one initially to an optimistic affirmation of the Western liberal belief in "history as progress."

Our study is the product of ongoing research by two political scientists who hope to contribute a treatment of issues of human rights, complementing that already provided by international legal specialists and/or by activists and journalists. It is our explicit assumption that the research technology of the social sciences is relevant to the substantive issues of human rights and can make a modest but unique contribution to the enhancement of human rights. More specifically, we intend here to continue addressing ourselves to a problem previously explored: the question of appraising the impact of a variety of international non-governmental organizations (NGOs) devoted to human rights (Scoble and Wiseberg,

1974, 1976a, 1976b). We must face two transitional questions: Why select these NGOs to study, out of all of the actors involved in the human rights arena? And to what extent do such groups, devoted exclusively to human rights, differ from the modal type of "international scientific and professional association" (ISPA)?

Before we can turn to such an analysis, we must briefly categorize the significant actors on human rights issues; and such a consideration ought logically to include both violators and protectors.

HUMAN RIGHTS: VIOLATORS AND PROTECTORS

From the perspective of a corporate actor frame of analysis, the major threats to human rights are posed by national governments, multinational corporations, and terrorist groups.

Whether acting on their own initiative or on behalf of a dominant social class, ethnic or religious community, and so on, state authorities (i.e., governments) have historically been first and foremost among the violators of human rights. Their power is made manifest through internal security forces (e.g., police, judges, state prosecutors, prison officials), and other civil servants. National governments, even those most vocal in their dedication to democratic freedoms, have all too readily fallen prey to the temptation to perceive "national security interests" ubiquitously, and as uniformly justifying repressive intervention in domestic conflicts. Furthermore, as nation-states have become effectively centralized in governmental form, as the executive has everywhere assumed dominance over the legislative, and as the technology for political repression has become more sophisticated, it has become possible for regimes to become fully totalitarian—not merely authoritarian; and the power of the state has become awesome in confrontation with the powerlessness of the individual.

The rights of individuals are also threatened by the increased salience of multinational corporations (MNCs), of which some 60% are based in the United States. The problems posed by MNC's are of particular relevance to Americans because all projections into the future (e.g., Perlmutter, 1968; also Hymer, 1972) assume that three-fifths of major MNCs will remain distinctively American enterprises—although with indigenous national managers to ensure local political protective coloration, to be sure. But, whatever the national control of the MNCs, for them "the bottom line" is indeed the bottom line of a very conventional profit/loss statement (the balance sheet).

Illustrations abound. In 1971 alone, Gulf Oil admittedly contributed more than $3 million to President Park and his political party, thereby contributing to Park's narrow reelection that year–and to the 1972 Constitution which institutionalized the present military dictatorship in South Korea. In this same period Litton, Exxon, and other major U.S. corporations entered into what they assumed would be mutually rewarding developmental-project contracts with the regime of the Greek Colonels. ITT's offer of $1 million to the CIA to prevent or undermine the Allende regime in Chile is now well-documented history. Perhaps less well-known is the fact that U.S.-based (and other) MNCs had been pressuring the Pinochet dictatorship to make Chile a test case in order to render the Andean Common Market harmless to Western capitalist interests–and, in particular, to employ Chile to undo Decision 24 of the Andean Pact (which decision, *inter alia,* set maximum limits on profit rates, on the repatriation of profits, and on nonnational ownership of the common stock of key corporations). And, in October of 1976, the MNCs were successful: Chile pulled out of the Andean Pact (de Onis, 1976).

In short, there is ample evidence that the major owners and top managers of these Euro-American corporations are truly one-dimensional people, fixated upon profit and growth before which all else, including human rights, must give way. Furthermore, in 1975, the Trilateral Commission released its report "On The Governability of Democracies." This Commission is the creation of David Rockefeller as much as it is of any single individual; it operates as a high-level Western European, Japanese, and American "policy idea discussion group" (much like the Committee on Economic Development or the Council on Foreign Relations within the domestic U.S. political system); it is made up of representatives of the leading MNCs, of their political elites, and of prestigious scholars from relevant social science disciplines. Although there was some internal dissent at the time, it is probably safe to say that the majority of the Trilateral Commission agree with Professor Huntington's (1975) prescriptions in that 1975 Report that economic well-being requires the sacrifice, or at least diminution, of political rights–even in the advanced democracies themselves. The anti-human rights behavior of MNCs should occasion no surprise: earlier, in the domestic political struggle over civil rights and race relations, major corporations (e.g., U.S. Steel), when requested, consistently refused to endorse the methods or the goals of the civil rights activists–they insisted that they existed only for business and to make profit but thereby contributed their potential influence to those who resisted any change.[1]

Terrorist groups have always existed; but the threat they pose is now exponentially magnified by advances in technology, particularly nuclear weaponry. Their actions jeopardize human rights both directly and indirectly—if and when their purpose is to panic the existing national political elite to blind reaction and undifferentiated repression of the populace. Even when the goals of terrorists are more limited—as they seem to be in most skyjackings (i.e., kidnappings to obtain hostages as a resource for political bargaining)—the threat they pose to human rights may still be large. This is so for two reasons. First, the potential targets of skyjackings are generally from the most affluent and powerful strata of their societies, even though they represent a very small proportion of the total population: they are thus in a position to demand and expect governmental responses to assure their safety. Perhaps more important, such terrorism may instigate a governmental response which is foolishly counterproductive and/or may provide the government with a convenient excuse for repression of dissidents within the domestic political system. An example of the latter is found in Section 88a of the West German Penal Code. This Section was revised in mid-1976 and designed to prevent the publication and sale of books that provoke violence and encourage acts of terror and crime. Everything from detective novels to historical studies could be banned by a fearful state judiciary giving wide powers to the police. And, indeed, one case has emerged already: *Haymarket*[2] was confiscated by the police in Landshut, Bavaria, on the grounds that the description of such historical events could inspire contemporary terrorists in West Germany.

The protectors of human rights can be categorized as national governments, intergovernmental organizations, and non-governmental interest groups.

Governments, which we have already characterized as the main offenders, *can* nevertheless be key actors for human rights, because, in the final analysis, given the persistence of the nation-state system, it is through national legislation and enforcement that human rights must be upheld. Moreover, international law is primarily intergovernmental law: there is no international government with direct jurisdiction over individuals (with the limited exception of the partial supranational authority of the European Economic Community); and it is governments which enforce (or fail to enforce) international standards. From a strategic point of view, and despite the optimistic indicators previously documented, there is little historical evidence to suggest that governments have been or will be the prime initiators of pro-human rights actions. Instead, the historical record is one of governmental inertia and indifference to human rights violations wherever they occur; in the exceptional instance, governments have been prepared to act to enforce human rights standards and then only under severely limited conditions. If governments are confident that no bad precedents will be set (permitting external interference with their own "domestic jurisdiction"), if they perceive a political (e.g., Cold War propaganda) or economic advantage, if there are no perceived costs to themselves or their allies, and if there is some domestic catalyst group demanding at least symbolic action, *then* governments may take a pro-human rights stand.

Since 1945, there has been an explosion in the population of intergovernmental organizations (IGOs), many of which are at least tangentially concerned with human rights. The point we stress here is that most of the efforts of such IGOs have been directed toward the drafting of new conventions and covenants on human rights. Without cynicism, we hold that the prime impact of such IGOs has been rhetoric and paper and perhaps a minimal consensual articulation of international norms (which constitutes a serial enumeration of potentially conflicting "human rights"). The reasons for this all but verbal paralysis are well documented: lack of power over individuals, lack of budgets, and lack of political consensus (alternatively, political conflict between member states). For these reasons, the major hope for humanizing the international order must come from the third category of actors: non-governmental organizations (NGOs).

HUMAN RIGHTS NGOs and ISPAs

Historically, non-governmental organizations have initiated whatever positive changes have occurred with regard to the status of human rights. In a sense, it has been the NGOs which have proposed and governments which have disposed. For example, the International Committee of the Red Cross and the Red Cross Movement have been the main movers in the codification of the Geneva Conventions providing battlefront treatment of the wounded, protection of prisoners of war, and at least minimal guardianship for non-combatants (Forsythe, 1975). The antislavery and women suffragette movements both further illustrate this proposition.

Although they are very few in number and very weak in conventional political resources, human rights NGOs—such as Amnesty International (AI) the International League for Human Rights (ILHR, formerly the International League for the Rights of Man), the International Commission of Jurists (ICJ), P.E.N. International, the International Committee of the Red Cross (ICRC)—are the focus of our present research, because they are exclusively and totally devoted to the furtherance of international human rights. The universe of NGOs from which this small "sample" is drawn is exceedingly large; there are several hundred NGOs which have consultative status with the United Nations, its Specialized Agencies, and regional intergovernmental organizations; and these are often far more endowed in political resources (such as membership and money) than are the human rights NGOs. But we accord relevance to this larger universe only when its units are involved–infrequently and tangentially–in human rights issues. When one examines the entire NGO populations, one finds that the largest single category is that of international scientific and professional associations. The human rights NGOs share certain common characteristics with ISPAs, but they also differ in certain fundamental attributes from them.

In common with ISPAs, the human rights NGOs are non-governmental; they are international (in membership, targets, and/or tactics); they are voluntary organizations in the language of sociologists; and they are "elitist" in the sense that their membership is drawn from middle and upper social classes; i.e., individuals who possess the skills and the discretionary income and time to devote to international politics (Scoble and Wiseberg, 1976b). In certain respects, the two are both products of the modern economies and societies of the Western world. However, a fundamental difference between human rights NGOs and ISPAs lies in the fact that the membership base of the former is not one of common profession or scientific expertise; rather it is one of shared ideology–a secularized political philosophy derived from Judeo-Christian principles. (This is not to say that the shared ideology may not disproportionately emerge out of, or be associated with, some occupations rather than others; for example, social science professors rather than commercial trust bankers. Nonetheless, it is shared ideological interest that forms the basis of interaction even where the human rights interest group seems occupationally derived, such as P.E.N. International or the International Commission of Jurists.) As a consequence of this ideological base, the human rights NGOs are first and foremost involved in human rights conflicts, whereas most ISPAs are drawn into such political conflicts only secondarily. That is to say, when an ISPA takes a stand on a human rights issue, it is almost invariably the case that a human rights NGO has acted as a catalyst, drawing the attention of members of the ISPA to a threat to their immediate occupational interest and, thereby, coopting the latter as a co-interest group in an ad hoc coalition centered on this one particular conflict.

Indeed, human rights NGOs often have difficulty in catalyzing ISPAs–even where they should logically be predisposed to act in defense of their members' interest–because most professional and scientific organizations fear political conflict. They avoid "sensitive" issues; because they are afraid of "politicizing" the organization, they have an inherent tendency to support the status quo. Thus, Schmidhauser and Berg (1971), tracing the development of the American Bar Association, have noted the "understandable ambivalence" of its leadership with respect to matters of public policy: "When a subject did not meet their approval, recommendations were dismissed on the ground that the problem was 'political.' Where the recommendations were considered ideologically sound, consideration was urged as essential to professional responsibility."

By contrast, from our perspective, human rights NGOs are, and of necessity must be, constantly politicized. We do not use "political" here in the artifically limited sense employed by most Americans, i.e., of electoral partisanship. Rather, human rights NGOs are centrally involved in struggles over power and its uses, although these groups do not themselves seek power. Put differently, they do not seek to become the authorities although they most certainly seek to influence them.

RESOURCE POTENTIAL

In assessing the influence of an NGO, or of any political interest group, it is helpful to conceive of a two-stage process, the first stage of which is directed to efficient prepolitical organization (which we term zero-order group activities because they are primarily "resource generating" in nature), while the second stage includes all of those group activities that are manifestly political, in the broad conceptualization stated above (and which we term first-order activities because they are primarily of a "resource applying" nature). In this section, we wish to concern ourselves only with the zero-order activities, which most social scientists and almost all Americans associate with the uniquely American value of bigness: increase in membership, increase in annual budget, increase in number of group projects undertaken, and so forth.

With regard to the zero-order indicators of growth, the specific human rights NGOs which we have studied all exhibit a rapidly increasing *potential* for effective political action. For example, Amnesty International, since its origin in 1961, has grown from a tiny handful of interested individuals to some 60,000 spread over all six continents. The International League for Human Rights, which had a budget of only about \$15,000 during the 1940s and 1950s, now has an annual operating budget of \$50,000. The International Commission of Jurists (which is not a membership organization as such, having a directive board of 40 internationally respected jurists and lawyers) has grown to a situation where it has support sections in the legal communities of 50 nations. P.E.N. International, founded in 1921, has more than 80 centers in more than 60 countries. The current budget of Amnesty International is in excess of \$500,000, while the combined budget of the International Committee for the Red Cross has grown to an annual minimum of \$15 million in the 1970s.

Any more refined statistical analysis would, however, be relatively meaningless: when a new political interest group appears on the scene, simple increase from "practically nothing" to "something" necessarily exhibits a very high *rate* of growth; therefore, quantitative indicators make sense only longitudinally and in the comparative examination of mature groups. Meanwhile, though, it should be clear from the statistical information cited above that no one of these human rights NGOs commands either the mass membership or the material wealth—the two basic resources in American politics—which are both cause and effect of the "political clout" of such established U.S. pressure groups as the AFL-CIO, the American Medical Association, or the American Bankers Association.

The real influence of these groups rests upon a combination of less tangible resources: the Lasswellian values of enlightenment, skills, respect and rectitude. These combine to produce solid, credible, uncompromised research on the fundamentally moral values embedded in that category of political conflicts dealing with human rights. These organizations have initiated research on the state of human rights which few national governments have exhibited either the willingness or ability to undertake.

Such research is in itself "influence" because, while there may not yet be a "world public opinion," nonetheless national political elites act as if one exists. That is to say, there is more than ample evidence to support the proposition that national elites strive mightily to conceal evidences of domestic repression (Kutner, 1970); they act in a manner we would term "embarrassed" when inevitably that evidence leaks to an external audience; and they grudgingly enact piecemeal policy concessions that are clearly intended to placate both external monitors as well as domestic opposition forces.

Information, to have any political effect, must be communicated. Although the human rights NGOs do not have any "mass" or external communications media, the compelling nature of the moral issues that they focus on, as well as their meticulous documentation, means that the serious media will convey the substantive content of their reports. And because the quality media and their own attentive publics pay attention to such human rights conflicts, other members of national political elites must pay attention as well. The communication may not involve a direct persuasion process or moral suasion. Moreover, since these NGOs are almost exclusively Western based, the initial purpose of communication is to convince Western elites and Western attentive publics; in fact, non-Western elites may not, and indeed often do not, accord legitimacy to the NGOs and their information. But such non-Western elites do respond to the more conventional leverage (cut-off of aid, denial of military purchases, restrictions on trade, and so forth) which Western governments can manipulate in world politics. Thus, research and communication are the first stage in what is often a highly indirect process of political interaction leading to impact on the policy-making process.

IMPACT: CONCEPTUALIZATION AND ASSESSMENT

An initial conceptual problem in our research is the operationalization of "impact." As we reviewed past research by political scientists, it became evident that the overwhelming proportion of that research has been quite situational-specific with regard to place and time. That is to say, most of the accumulated research operationalizes impact in terms of winning or losing in conflicts arising just prior to and at discrete decision points in formal governmental institutions. This is as true for domestic political research, such as that dealing with "Congress passing a law" or "community power structure," as it is for research on international politics. In the latter case, for example, measures of impact have been quite narrowly confined to general policy formulation in formal governmental institutions and, more recently, with increasing attention being paid to law implementation and law adjudication (whether one conceives of these as bureaucratic problems at the intergovernmental level or problems of national implementation of international standards).

However, we find it useful to conceive of "impact" in a second sense as well, even admitting that this less situationally specific meaning is both long-term and nebulous. This second usage of impact is in the realm of social-psychology—and by that we mean the long-term cumulative impact of this small subset of NGOs in defining and expanding attentive publics in various nations which not only consume information about human rights but are willing to act on it. (The one group of scholars which has concerned itself with a variant of this second sense of impact has been international legal scholars—specifically those who address the question of whether the massive volume of UN covenants, declarations, and resolutions can be, and should be, considered "law" rather than simply aspirations, even though such law is not self-executing. The problem with this legal approach is that it has remained largely a high-level abstract philosophical discussion without reference to the political forces requisite to making moral principles politically salient.)

In moving from conceptualization to assessment, and initially utilizing the more common meaning of impact with reference to law-making, most human rights NGOs have been concerned with one or more major legislative efforts: the ILHR was deeply involved in the efforts to formulate the 1948 Universal Declaration of Human Rights and subsequent covenants; the ICJ, consistent with its dedication to "the support and advancement of those principles of justice which constitute the basis of Rule of Law," has been particularly active in the drafting of international covenants furthering this principle and with such efforts as the Draft Convention on Territorial Asylum and the United Nations Standard Minimum Rules for the Treatment of Prisoners; AI was very active in the efforts (which it, in fact, initiated in 1972) to obtain the 1975 international declaration of the UN General Assembly condemning torture as a crime against humanity; and P.E.N. International has been much concerned for the protection of the rights of individual authors in the recent revision of the International Copyright Law. In general, each of these organizations—as well as many other NGOs—either has felt sufficiently efficacious and rewarded to participate in other general law-making efforts, or has acted out of fear that nonparticipation would mean that its interests would go totally unrepresented.

However, international law-making, through the drafting of treaties and the codification of norms in the declarations of intergovernmental organizations, is only one aspect of international legislative processes. Equally significant have been the efforts of these groups to bring national law into conformity with international law by the ratification of covenants and the modification of domestic law. In this second meaning of implementation, human rights interest groups have been far less successful, as indicated by the small number of senatorial ratifications of pending conventions that the U.S. Executive Branch has helped draft, and has voted for, in international forums.[3] For example, the Convention on the Prevention and Punishment of the Crime of Genocide, which was approved in the General Assembly in 1948 and entered into force in 1951, remains to be ratified by the United States (see Harvard International Law Journal, 1975).

Within the American political system, the explanation for this relative lack of impact is not hard to deduce: it is relatively costless for the U.S. Executive Branch to cooperate in the drafting and successful adoption of legislation that would significantly advance human rights—and give the United States a good public image, especially when contrasted with that of its ideological opponent, the USSR—in such bodies as the UN General Assembly. At the same time, the executive engages in only a pro forma introduction of these same materials for ratification by the U.S. Senate because it does not wish to risk alienating powerful national groupings. Indeed, where powerful interest groups mobilize in opposition to such ratifications, the executive branch has rapidly retreated. At least, this has been the case from the early 1950s until the Carter Administration took office and President Carter pledged himself to gaining Senate ratification of the International Covenant on Political and Civil Rights, The International Covenant on Economic, Social and Cultural Rights, and the Genocide Convention. Whether this pledge was more than rhetoric, and whether the administration was willing and able to generate enough pressure to overcome whatever Senatorial opposition was mobilized against the ratifications, remained to be seen at the time of this study. But, in this regard, it is instructive to examine the two decades of executive inertia and disinterest.

Thus, it is relevant to recall that, in the first years after World War II, the United States took the lead in the United Nations, its Commissions and Specialized Agencies, in working for the codification of human rights legislation. Moreover, within the American political system, there were some efforts manifested to bring U.S. domestic law into line with international human rights standards. The most notable, in this regard, was the opinion of an intermediate California court, in the case of *Sei Fujii v. State of California* (1950), which struck down California's Alien Land Law (forbidding non-U.S. citizens from owning land in California) on the grounds that it violated human rights provisions of the United Nations Charter; the Charter, being a treaty entered into by the United States, was equal under the Supremacy Clause (art. VI, clause 2, U.S. Const.) to a federal statute and thus superseded any inconsistent state legislation (Buergenthal, 1974). Although this was subsequently reversed on appeal to the California Supreme Court, nonetheless, it immediately resurrected fears of the "conservative Republican-Southern Democratic" majority coalition which had dominated the U.S. Congress since 1938, and of the conservative-controlled American Bar Association (Schmidhauser and Berg, 1971), that the *Missouri v. Holland* precedent of 1920 would now be expanded by an internationalist President and an activist U.S. Supreme Court: that through treaties, and the Supremacy Clause once again, foreign-made human rights law could now bind state and federal judges—even at the expense of domestic law to protect private property. Thus arose the battle over the Bricker Amendment to restrict the treaty-making powers of the President and to make explicit what had been previously only implicit, that the Constitution limited international agreements just as it did the powers of the federal government.

The Bricker Amendment—an amendment to the U.S. Constitution proposed by the Republican lawyer from Ohio, Senator John Bricker, in at least seven different forms and wordings in his own name, and with alternates and substitutes agreed in this era—had as its main thrust the statement that "a provision of a treaty which conflicts with this Constitution shall not be of any force or effect" (U.S. Senate, 1954). While, from an historical perspective, the Bricker Amendment was one detailed part of a sustained attack on the office and powers of what later came to be termed "The Imperial Presidency," and was, more narrowly, a direct legal assault on the practices of Presidents and Secretaries of State (especially during Franklin D. Roosevelt's tenure) to employ "executive-agreements" in order to achieve some foreign policy purposes that should logically and legally have required the slower, more cumbersome and "chancy" treaty-making process itself, its implications for treaty ratification were equally great. The Eisenhower-Dulles Administration decided, therefore, to strike a bargain under the concerted pressure from Senator Bricker, the American Bar Association, and other conservative forces: in 1953, Secretary of State John Foster Dulles announced that the United States would not sign the Convention on the Political Rights of Women, would not press for Senate approval of the Genocide Convention or any other international human rights treaty, and conveyed the import of this decision to the United Nations (Van Dyke, 1970: 131). Thus, the ratification of human rights treaties was shelved for two decades.

From a different perspective, a part of the explanation for the enormous influence of Bricker and the American Bar Association in this arena is the fact that there has been no effective domestic constituency for human rights. Moreover, given this lack of concern, "in the Department of State international human rights matters (fell) into a bureaucratic no-man's land" (Buergenthal, 1974: 616), and there was only one State Department official—in the International Organization section—with responsibilities for human rights.

Furthermore, when one looks beyond the formal law-making stage to those of implementation and adjudication, it becomes evident that, until recently—the last decade or so—most non-governmental actors have been naive in accepting the passage of general law as a definitive and final victory. The IGOs, such as the UN Human Rights Commission, have been equally naive: for example, it was not until 1967 that the Commission reversed its own 1947 ruling that it had no power to take action on complaints concerning human rights violations, and only in 1971 that the Commission expanded its mandate to deal with gross violations of human rights generally—until these changes, it devoted almost all of its budget, time, and personnel to law drafting, paying scant attention to the problem of implementation (Wiseberg and Scoble, 1977). It may be that the Human Rights Commission and the human rights NGOs have more impact than we have credited them with here; but that is more properly an evaluation of their long-run sociopsychological effect, a question we will treat in the next section.

EVALUATING THE INFLUENCE OF NGOs

In attempting to evaluate the political influence of an interest group, the analyst should be aware of twin dangers. The first risk is that of attributing influence where none in fact is demonstrated; for example, an error easily arrived at in political science given its long history of detailed descriptive concern for the *activities* of political interest groups. From such (presumably accurate) description, the imaginative leap of inference is made. This is probably the more common of the twin dangers precisely because it seems the case that those who research and write on the subject of human rights (a) tend to believe that human rights *are* important values, and (b) want to believe that organizations that proclaim their devotion to such values are effective as well as sincere.

Alternatively, there is the danger—one which blossoms in the offices of Social Science Buildings—that the demand for scientific rigor, for quasi-experimental proof, is so strict as to deny any attribution of influence to the actions of the group under study, when some influence did in fact exist.

The above suggests that in examining the group actors in human rights conflicts one cannot approximate Campbell's quasi-experimental design (Campbell and Stanley, 1966), for there are too many uncontrolled, and uncontrollable, variables. One cannot assess in a systematic, rigorous, quantitative fashion the direct policy impact of a single human rights NGO for three recurring reasons: first, human rights conflicts invariably reveal two or more such NGOs working in the same direction, for the same policy; second, there is no visible organized opposition in such "conflicts" simply because legitimacy is monopolized by the "pro" side; and, finally, the researcher can never have psychological access to the authoritative actors, can never get into their minds to determine what variables they considered and how they weighted them.

In addition, there is a further broad problem. The researcher cannot logically focus on a single group *in vacuo,* but must also examine that group in the context of the system in which it exists. Using some of the conventional indicators of influence, and focusing upon but a single group, the researcher might initially judge that group as highly influential. But if the frame of reference is expanded to include all of those groups and institutions with which that first group is interacting, it may well be observed that the world political situation has merely been transformed from a "slack" to a "taut" situation (where the slack-taut continuum refers to the potential resources available, the proportion of these devoted to politics, and the efficiency with which these are exploited).[4] That is to say, progress in defending human rights has equally and oppositely been matched by "progress" in the capacity of states to violate precisely those rights. For the purposes of illustration, Amnesty International can now claim the release of more than 8,000 prisoners of conscience since the organization was formed in 1961; but how is one to evaluate this information, given the vast rate of increase (also since 1961) in the political prisoner population the world over?

It should be clear that there is no single definitive measure of impact. The research must rely on a "multitrait-multimethod" approach in which most "metrics" will actually be non-quantitative (since there do not exist good quantitive indicators of impact). Thus it is appropriate to return to our earlier conceptualization of impact in terms of long-term sociopsychological change in values, attitudes, and opinions concerning human rights.

One can make a tentative evaluation that human rights NGOs have had a cumulative impact on American opinion elites and decision makers: this is evident in the response to the appeal of these organizations for American membership, contribution, and participation, and also in the recent responsiveness of the relevant committees of the Congress. However, it is too soon to judge whether international human rights is now permanently on the domestic agenda of the United States and of the other advanced nations.

EMERGING AND ALTERNATIVE STRATEGIES

To consider the activities of human rights NGOs from a somewhat different perspective, they have been engaged in a threefold strategy. The first, and clearly an instrumental precondition for the other two, has been coalition-building to maximize effectiveness. The second has been intervention in specific crisis situations. And, lastly, the third strategy has been directed towards the strengthening of existing institutions and the building of new institutions for world order.

With respect to coalition-building, because the human rights NGOs recognize that they are few in number and have limited resources, they realize that they must engage in cooperative action not only with each other (a fact facilitated by overlapping membership coupled with specialization of function) but also to draw in peripheral actors. To the extent that they have a goal of building a mass citizen movement on behalf of human rights, the logical target of such a strategy is not isolated individuals but instead people already organized in groups: the prime candidates here are trade unions, church groups, and professional and scientific organizations (both national and international).

Domestically, an informal coalition has been in existence for several years among Amnesty International, the American Association for the ICJ, and the International League for Human Rights. These three have engaged in cooperative informational activities with the Protestant and Roman Catholic Churches—especially with regard to the four national human rights crises of Southern Africa, Brazil, Chile, and now Argentina. On the other hand, a second and distinctly different coalition has recently formed from the residual legatees of the peace/Vietnam protest movement—one which aggressively engages in legislative lobbying in Washington—the Coalition for a New Foreign and Military Policy[5] and its Human Rights Working Group. The Coalition, with several full time staff members in Washington, is an umbrella organization of organizations the most active of which are either church-affiliated (e.g., Clergy and Laity Concerned, American Friends Service Committee, National Council of Churches) or issue/area specific (e.g., Washington Office on Africa, Washington Office on Latin America, Indochina Resource Center, TAPOL—Campaign for the Release of Indonesian Political Prisoners); and it engages in a very wide array of issues.

It would, however, be premature to predict a coalescence of the old line human rights NGOs and this particular Coalition into a single grand coalition similar to the Leadership Conference of the Civil Rights battles fifteen years ago. For one thing, the human rights NGOs (which either have or are seeking tax-exempt status) have strictures against direct lobbying for specific legislation; for another, they wish to avoid dilution of members' interests and other resources by focusing on too wide a range of issues; finally, they tend to view human rights as non-ideological and are concerned to avoid contamination of this neutral stance with what they perceive as the left-ideological position of the Coalition.

Further, on coalition-building, one should note a recent tendency for a number of scientific and professional associations, at the national level and in combination, to take stands on certain issues of political repression and, in some instances, to create new standing committees with an obvious anticipation that these human rights issues will continue into the foreseeable future. To illustrate, in 1976 the National Academy of Sciences formed a new Human Rights Committee, which has sought to pressure Argentina, Uruguay, and the Soviet Union and to force these regimes to reveal the whereabouts and to vouch for the safe treatment of scientists imprisoned for political reasons; the American Mathematical Society recently joined with similar societies in Canada, Great Britain, and France for a collective protest of the arrest and prison treatment of a prominent Uruguayan mathematician; and the American Association for the Advancement of Science in 1976 formed a new committee on scientific freedom and responsibility which is to function as a "human rights clearing house" for the three hundred scientific and technical associations affiliated with the AAAS and which has already lodged a protest with the Organization of American States concerning Argentina's treatment of scientists in that troubled country (Scully, 1977).

We do not know whether the human rights NGOs are directly responsible for this partial politicization of the national scientific and professional associations. One can state, however, that in the performance of their primary function of data collection and publicity on political repression, the human rights NGOs contributed to the new political awareness of the members of such scientific and professional associations. Further, one can state that the evidence thus far confirms our prediction about the limited nature of involvement in human rights issues of ISPAs: that is, to the extent that such associations respond to human rights issues, the response is based upon direct threats to their membership or to colleagues in other nations. Moreover, while in some respects these new ventures would seem to contradict our earlier hypothesis—that ISPAs shun the kind of politicization that the human rights NGOs are necessarily involved in, the evidence is hardly strong enough at this point in time. We would, however, point out that in each of the above-recounted instances, the organizational response has been the same—limited and conflict-avoiding; that is, a special committee has been set up to permit those members who feel they must act to act, at the same time that the formal leadership and the full membership avoid taking a stand and politicizing the organization.

For these reasons, the prospects for coalition-building with ISPAs by the human rights NGOs are only limited at this time. That is, it will probably require building new ad hoc coalitions as different crisis issues emerge involving specific scientists and/or professionals.

As to the second major strategy of human rights NGOs, these groups will continue to respond to gross violations of human rights wherever they occur. Obviously, their ideal is to have an early-warning system sufficiently developed so that there no longer would be "crises," which implies being in a passive condition and having to react to the initiative of others. But how one achieves this ideal state remains elusive in the world today.

Therefore, the long-term goal of these groups lies in the third strategy, that of institutional strengthening or creation with the hope of either obviating these periodic crises or, at least, of having on hand intergovernmental institutions adequate to the task of minimizing and ameliorating such situations. In this context, it is relevant to note the efforts of human rights NGOs directed toward the creation of a United Nations High Commissioner on Human Rights, just as we might anticipate that the NGOs will seek to take the recently established human rights officers in the U.S. Department of State and extend these downward through the bureaucratic structure to the embassy level.

CONCLUSION

One of the most dramatic recent changes in human rights issues has been affirmative action on human rights in the U.S. Congress. Major credit for the strong pro-human rights stand of Congress must go to Representative Donald M. Fraser, who in 1973 chaired a series of human rights hearings in the U.S. House Subcommittee on International Organizations (1974)—the first comprehensive hearings that examined the human rights situation throughout the world and that strongly criticized the administration's indifference to human rights violations. Congressional concern first manifested itself in a very tentative one sentence human rights provision in the Foreign Assistance Act of 1973 and, in the face of strong Administrative resistance, in a somewhat stronger provision in 1974. The first major positive action was, however, in 1975 in the form of the floor amendment by Representative Harkin to the Foreign Aid Bill: this amendment prohibited U.S. aid to "any country which engages in a consistent pattern of gross violations of internationally recognized human rights," unless such assistance "will directly benefit the needy people in such country." While the Harkin Amendment applied essentially to American bilateral aid, the essential language of Sec. 116 was attached to the Inter-American Development Bank and the African Development Fund Bill (HR9127) on December 9, 1975 (89 Stat. 860, Sect. 310, 1975).

The second major initiative was the Fraser Amendment—Section 502B of the International Security Assistance and Arms Control Act of 1976, which added a human rights amendment to the 1961 Foreign Assistance Act. Signed into law by the President on June 30, 1976, Section 502B declares "a principal goal of the foreign policy of the United States is to promote the increased observance of internationally recognized human rights," compels the Administration to make "a full and complete report" on the human rights situation in every country receiving security assistance, permits the foreign relations committee of either House or Senate to demand further information on human rights in aid-recipient countries, and provides for the termination or restriction of security assistance "to any country which engages in a consistent pattern of gross violations of internationally recognized human rights (22 U.S.C.A. Sect. 2304, Supp. I, 1976). It was under the Harkin Amendment that aid was first denied to the Pinochet military regime in Chile and, subsequently, to Uruguay. Then, in March 1977, Secretary of State Vance announced that foreign military sales credits were to be reduced for Ethiopia, Argentina, and

Uruguay on human rights grounds, under Section 502B (New York Times, March 1, 1977).

Coincident with these public manifestations of concern for human rights, the Division of Overseas Ministries of the National Council of Churches of Christ in the U.S.A. completed an internal reorganization which, in early 1977, resulted in the creation of a Human Rights Office. And in the general euphoria of the apparent new policy, there have been a large number of conferences—both scholarly and activist-oriented—and other activities centered on human rights.

However, not everything on the human rights scene can be construed as progress. In mid-1977, the International League for Human Rights and other human rights NGOs are concerned about a possible concerted attack on them by a new (and strange) coalition—led by the right-wing military dictatorships of Argentina and Chile and the Communist regimes of the USSR and the Ukraine—when their consultative status with the ECOSOC will come up for periodic review in 1978. Their fears concerning a sustained effort to decertify or disbar them from consultative status in the United Nations system may be exaggerated, but it is clear that these NGOs are perceived as a threat by many member states of the United Nations that resent their outspoken criticism about human rights violations in their countries (see Wiseberg and Scoble, 1977). In 1974-1975, for example, after the NGOs broke the unwritten rule of the United Nations Commission on Human Rights (1975) about naming the names of countries violating human rights, the ECOSOC issued a strong statement of censure:

> Having been made aware by the Commission . . . that some non-governmental organizations have occasionally failed to observe the requirements of confidentiality . . . and having been made aware further that the oral interventions of some non-governmental organizations on matters affecting Member States have often shown disregard for proper discretion . . . (the ECOSOC) decides that any non-governmental organization failing to show proper discretion in an oral or written statement may render itself subject to suspension of its consultative status.

Additionally, the removal of the Human Rights Division of the United Nations from New York to Geneva and the decision that all meetings of the UN Commission on Human Rights and its Subcommission on the Prevention of Discrimination and the Protection of Minorities shall henceforth be held only in Geneva must be interpreted both as a move by member states of the United Nations to put human rights on a back burner and to reduce the influence of human rights NGOs in these bodies. Finally, one might note the recent suggestion by the Soviet Union to dismantle the Subcommission on the Prevention of Discrimination and the Protection of Minorities–the UN human rights body upon which persons serve as individuals rather than as state representatives–supposedly to effect a rationalization of the United Nations system in line with recommendations of the Group of Experts on the United Nations structure. This is clearly a ploy to recapture control over human rights implementation processes for sovereign nation states and to lessen the influence of individuals and NGOs.[6]

If the human rights NGOs truly seek positive change with regard to the human condition the world over, if they seek an even-handed human rights policy–universal criteria uniformly applied to the behavior of our allies as well as our enemies–then now is precisely the time to intensify the pressures (however discreetly) upon the President, the Congress, other nation-states, and various IGOs, such as the UN Commission on Human Rights.

NOTES

1. On U.S. Steel and the civil rights movement in Birmingham, Alabama–where in 1963 U.S. Steel was the largest employer, see Bachrach (1967).

2. This is an historical analysis of the 1886 socialist demonstration in support of an ongoing strike demanding adoption of the 8-hour day. The demonstration was in Chicago's Haymarket Square, through which the Chicago Police swept in an effort to break up the workingmen's rally. A bomb was thrown (by an anarchist, apparently) which wounded 70 policemen (seven of whom ultimately died); this caused a "police riot" in reaction, and the subsequent criminal trial of eight anarchists rounded up in a police dragnet. For further details, see Hofstadter and Wallace (1971).

3. The only human rights conventions, of the major nineteen existing instruments, thus far ratified by the U.S. government are the 1926 Slavery Convention; its Protocol; the Supplementary Convention on the Abolition of Slavery and the Slave Trade; the Protocol Relating to the Status of Refugees; and the Convention on the Political Rights of Women (World Association of Lawyers, 1976).

4. Robert A. Dahl (1963) has conceptualized power and influence in terms of this threefold reference to potential political resources.

5. The Coalition for a New Foreign and Military Policy grew out of the merger of two earlier coalitions–the ad hoc Coalition for a New Foreign Policy and the Coalition on National Priorities and Military Policy. It is formally a coalition of 33 organizations, though it cooperates with many other organizations and its task force–the Human Rights Working Group–has attracted the participation of many organizational representatives of groups which do not feel that they can formally affiliate.

6. This is suggested by Bernard D. Nossiter (1977) of the Washington *Post* in his article, in which he reports the recent publication of a book by V. A. Kartashkin, *The International Defense of Human Rights,* that presents this argument. The assumption is that Kartashkin is articulating Soviet policy in this volume.

REFERENCES

BACHRACH, P. S. (1967) The Theory of Democratic Elitism: A Critique. Boston: Little, Brown.

BROWNLIE, I. [ed.] (1971) Basic Documents on Human Rights. Oxford, UK: Clarendon Press.

BUERGENTHAL, T. (1974) "International human rights: U.S. policy and priorities." Virginia J. of Int. Law. 14, 4: 611-621.

CAMPBELL, D. T. and J. C. STANLEY (1966) Experimental and Quasi-Experimental Designs for Research. Chicago: Rand McNally.

Commission on Human Rights (1975) Report on the Thirty-First Session, 3 February-7 March, 1975, E/5635 E/Cn. 4/1179.

DAHL, R. A. (1963) Modern Political Analysis. Englewood Cliffs, NJ: Prentice-Hall.

De ONIS, J. (1976) "Chile is pulling out of the Andean Pact." New York Times (October 7).

FORSYTHE, D. P. (1975) Present Role of the Red Cross in Protection. Background Paper 1, Joint Committee for the Reappraisal of the Role of the Red Cross. Geneva: Henry Dunant Institute.

Harvard International Law Journal (1975) 16, 3: 683-704.

HOFSTADTER, R. and M. WALLACE (1971) American Violence. New York: Vintage.

HUNTINGTON, S. P. (1975) "The United States," in M. J. Crozier et al. (eds.) The Crisis of Democracy. New York: New York Univ. Press.

HYMER, S. (1972) "The multinational corporation and the law of uneven development," in J. N. Bhagwati (ed.) Economics and World Order: From the 1970s to the 1990s. New York: Macmillan.

KUTNER, L. (1970) "World Habeas Corpus: ombudsman for mankind." University of Miami Law Rev. 24, 2 (Winter): 352-388.

NOSSITER, B. D. (1977) "Soviets foiled in effort to alter UN rights report." Washington Post (February 2).

PERLMUTTER, H. V. (1968) "Supergiant firms of the future." Wharton Q. (Winter): 8-14.

SCHMIDHAUSER, J. S. and L. L. BERG (1971) "The American Bar Association and the Human Rights Conventions: the political significance of private professional associations." Social Research 38 (Summer): 362-410.

SCULLY, M. G. (1977) "U.S. scholars step up human rights drives." Chronicle of Higher Education (May 9).

SCOBLE, H. M. and L. S. WISEBERG (1976a) "Amnesty International: evaluating effectiveness in the human rights arena." Intellect 105, 2377 (September/October): 79-82.

--- (1976b) "Human rights NGOs: notes toward comparative analysis." Human Rights J. 9, 4: 611-644.

--- (1974) "Human rights and Amnesty International." The Annals 413 (May): 11-26.

SOHN, L. B. and T. BUERGENTHAL [eds.] (1973) Basic Documents on International Protection of Human Rights. Indianapolis: Bobbs-Merrill.

U.S. House of Representatives Subcommittee on International Organizations (1974) International Protection of Human Rights: The Work of International Organizations and the Role of U.S. Foreign Policy. Hearings before the Subcommittee on International Organizations and Movements of the Committee on Foreign Affairs, House of Representatives, Ninety-Third Congress, First Session, 1973. Washington, DC: Government Printing Office.

U.S. Senate (1954) S.J. Res. 1, Sect. 1, Eighty-Third Congress, Second Session.

VAN DYKE, V. (1970) Human Rights, the United Nations and the World Community. New York: Oxford Univ. Press.

WEISSBRODT, D. (1977) Testimony before the Subcommittee on Foreign Assistance of the Senate Committee on Foreign Relations, March 4.

WISEBERG, L. S. and H. M. SCOBLE (1977) "The International League for Human Rights: the strategy of a human rights NGO." Georgia J. of Int. and Comparative Law 7 (Spring supplement): 231-287.

World Association of Lawyers, Committee on Human Rights (1976) The Ratification of International Human Rights Treaties. Washington, DC: Author.

COMMENTARY

James Frederick Green

This brief comment on the stimulating study by Professor Scoble and Professor Wiseberg will be limited to only a few aspects of the role of international nongovernmental organizations in the promotion and protection of human rights. Because the authors make a number of observations about United States policy in this field, the role of national organizations is also included.

The authors correctly note that much of the early work of the United Nations was devoted to the drafting of international standards of conduct—the Universal Declaration of Human Rights, the Genocide Convention, the two International Covenants, and the numerous declarations and conventions on women's rights, racial discrimination, labor, education, torture, and related subjects. They seem to underestimate the importance of that pioneer activity, for this was the first time in history that sovereign states accepted standards by which peoples can measure their governments. The authors also overlook the significant work of the International Labor Organization and of UNESCO in the formulation of conventions, declarations, and recommendations.

The Scoble-Wiseberg chapter similarly underestimates the role of international and national non-governmental organizations in the preparation of these instruments, through lobbying in their national capitals and in the corridors of international conferences. Public participation in such conferences is a relatively new phenomenon in diplomacy. The first example was probably the World Disarmament Conference at Geneva in February 1932, when a special plenary session was arranged for the representatives of citizen groups—peace societies, churches, women's groups, youth groups, and others—to present petitions urging disarmament. (The present author represented, as a college senior, American and British youth on that occasion.) In the Spring of 1945, the Department of State was swamped with requests from nongovernmental organizations to attend the newly announced United Nations Conference at San Francisco. As seating in the meeting halls and hotel space were both limited, the Department hit on the idea of accrediting two representatives from each of a small number of national organizations as "Consultants to the U.S. Delegation." The 42 organizations thus participating in the San Francisco Conference in 1945 did much to effect the inclusion of seven references to human rights and fundamental freedoms in the Charter—a subject that was omitted in the Dumbarton Oaks Proposals.

The demand of the NGO representatives and a number of delegations for an "International Bill of Rights," to be attached to the Charter, was unsuccessful, because of lack of time. Later, however, the nongovernmental organizations, especially the religious ones, took an active part in pressing for the inclusion of the Universal Declaration and the two International Covenants.

As the authors indicate, the current emphasis is on the observance and implementation of these agreed instruments, and on the role that nongovernmental organizations can play in that process. The Sub-Commission on the Prevention of Discrimination and the Protection of Minorities has been empowered to examine in closed sessions communications alleging violations of human rights and to recommend further study where a pattern of consistent and gross violations seems evident. So far, however, action has been so successfully blocked that no report has yet been submitted by the Commission on Human Rights to the Economic and Social Council.

Progress is also being made at present through another technique—the preparation of a study of some specific right, prepared by a rapporteur and based upon information provided not only by governments but also by nongovernmental organizations and other sources. Recent studies have included such subjects as rights of indigenous peoples, slavery and practices akin to slavery, migrant workers, and aliens in their state of residence. The slavery report named names: four Central American countries where forced labor was reported to exist; and Paraguay, where Indian children were reportedly sold for $2 a head.

The work of Amnesty International, the International Commission for Human Rights, and the other organizations discussed by Scoble and Wiseberg in exposing violations of human rights around the world is commendable. Indeed, given the nature of governments in a large majority of countries, that work is indispensable. It is governments that violate human rights or permit their violation; it is the press and the private sector that must protect the citizen against the government.

This work involves risks, however, for most of the members of the United Nations do not welcome exposures of their imperfections. To paraphrase Gilbert and Sullivan, "An NGO's life is not a happy one." For example, in the 1975 session of the Commission on Human Rights, the charge that human rights were being violated in a number of named countries made by representatives of Amnesty International and the World Conference of Religion Through Peace provoked considerable protest. As a result, noted in the Scoble-Wiseberg chapter, the Commission submitted a resolution to the Economic and Social Council recommending that nongovernmental organizations must comply with established procedures for written and oral submissions, and warning that any organization failing to show "proper discretion" may subject itself to suspension. The Council accepted the recommendation, but softened its tone somewhat by adding a preambular paragraph recognizing that "non-governmental organizations have an important role to play in the promotion and protection of human rights."

Three related questions about future action are provoked by the Scoble-Wiseberg study: What role should nongovernmental organizations play in the United Nations? What role should national nongovernmental organizations play in Washington? What should be done to strengthen nongovernmental organizations throughout the world?

First, it may be urged that international nongovernmental organizations maintain and, wherever possible, increase their pressure for exposure of violations of human rights everywhere. The General Assembly and the Commission on Human Rights spend much time in considering violations in three favorite target areas—Southern Africa, Chile, and Israeli-occupied territories—that contain only 50 million people. There are 4 billion people elsewhere in the world, however, whose rights also require attention. The UN organs are beginning to look at other areas, but they still tend to neglect the suppression of freedom in the communist countries and in much of their Third World, as well as limitations on freedom in some of the developed countries. Constant exposure of violations must be maintained and increased—through existing institutions, through new ones like the proposed High Commissioner for Human Rights mentioned by the authors, and through regional institutions. There is widespread opposition to the idea of a High Commissioner, but there are alternative techniques for creating an international ombudsman that could be pursued. Regional protection of rights is well developed in Western Europe and Latin America, but almost totally lacking in other regions. On both the global and the regional fronts, the nongovernmental organizations need to become more active.

Second, it may be suggested that nongovernmental organizations in the United States could do much more in pressuring the Executive Branch and the Congress to take action in the field of human rights. The record of the United States in signing and ratifying the many conventions concluded by the United Nations, the International Labor Organization, UNESCO, and the Organization of American States is hardly an admirable one. The leadership that the United States once exercised in international forums has vanished during the decade when the Government was preoccupied with Vietnam and Watergate. As the Scoble-Wiseberg study notes, however, that leadership was revived by the Carter Administration.

Third, the role of international and national nongovernmental organizations in influencing foreign policy in general and human rights policy in particular is minimal or nonexistent in much of the world. The international nongovernmental organizations and their national affiliates should devote much of their resources to keeping in touch with their local groups throughout the world and in helping organize local groups where they do not exist, especially in countries where human rights are in jeopardy.

COMMENTARY

Berhanykun Andemichael

Before commenting specifically on the individual chapters, it is necessary to distinguish first the different levels at which we may observe the existence or lack of social and cultural impact of ISPAs—the impact on their membership, on the general public in a particular society, and on the global society. A crucial question at all levels is: What effect do ISPAs have in reducing inequality and injustice and in promoting mutual cultural enrichment and development? At the global level, we live in one world only in a physical sense, but that at least three sociopolitical worlds are coexisting today: the developed Western world; the socialist developed world, and the less developed Third World. Little has been specifically said in these studies whether ISPAs have made any impact on the sociocultural relations of the East and the West—in facilitating detente and in actively promoting scientific, cultural, and social exchanges. Carr's study on "the international policy impact of teachers' organizations" gives an illuminating account of the positive impact on the drafting of the UN Charter, the establishment of UNESCO, and the lack of impact during the Berlin crises. But none of the chapters address themselves to the more recent period of detente which culminated in the Helsinki Conference on Security and Cooperation in Europe. While the main thrust of activity was based on governmental initiatives, the role of nongovernmental contacts in forums such as the Pugwash Conferences cannot be ignored. As regards the North/South confrontation, an aspect of the dependency nexus has been clearly stated by Angell, especially where he enumerates the elements of asymmetry among social scientists in the world due to the colonial heritage. But the impact of ISPAs and human rights NGOs on the fundamental regional inequalities, which take the form of racial or ethnic discrimination in some areas and institutionalized socioeconomic domination in others, have not been spelled out. The decolonization and anti-racism movements, for example, demonstrate the significant impact that NGOs, including many ISPAs, have had since the abolitionist movement on the worldwide structure of social injustice.

At the level of the general public in particular societies, a relevant issue is the tension in many ISPAs between the task of protecting and promoting the group values and interests and the task of providing public service. The contrast between the public-issue (human rights) NGOs discussed in the provocative Scoble-Wiseberg chapter, and the ISPAs examined in the Angell and Carr studies provides a helpful context in addressing these issues. These two chapters on ISPAs give considerable evidence about harmonious relationships between group interests and the public interest. But it would also be worthwhile to show any negative impact with examples from organizations facing a conflict of values in this regard as in the case of the World Medical Association and the national medical associations in countries like the United States.

At the level of the membership, the role of ISPAs in promoting a sense of fellowship and camaraderie within and across national boundaries has been amply described. The other side of the coin is the sense of elitism that often accompanies professional societies. In more developed societies, there is perhaps less concern about elitism as a problem than in developing societies. Unless the public service function of ISPAs in developing countries is firmly established, and the code of ethics is adjusted according to the development needs, ISPAs could serve to perpetuate the educational, economic and "cultural" gap between the small urban middle class and the poor masses. Little has been said in the studies about this basic issue.

As for specific comments, Angell's definition of ISPAs as fellowships of high competence "acquired by talent or training or both" and his distinction between the research orientation of scientific associations and service orientation of professional associations are most helpful in understanding the subject. The problem is in applying the definition to marginal cases. For example, would advocacy or public issue NGOs like the International Commission of Jurists, with a membership largely from a few closely related professions, be included as ISPAs? Perhaps they should be, because of the close relation between the professions concerned and the issues advocated (e.g., law and justice). The exclusion of organizations like the World Council of Churches and the International Committee of the Red Cross can be easily explained since they clearly belong to the category of service-giving voluntary agencies without being linked to a particular profession.

The description of the process of integration provides a clear idea of the specific effects of the various channels used. Of particular interest is the analysis of the congresses, working groups, journals, and so forth, as promoters of transnational ties, and the discussion of the trends toward internationalization in terms of subjects of concern, international scholarly inputs, and the sharing of knowledge with developing societies. But this analysis gives few insights into the overall impact. The point on the normative role of ISPAs is well taken, but less convincing is the remark about the ethical code of scientific associations that "the principles are well known among scientists in developed countries, but ISPAs bring them to the attention of scientists in developing countries." Scientists in developing countries are no less familiar with those codes, which are often internalized during their scientific training. Probably all scientists would benefit from a reiteration of ethical codes through world journals. An even more basic point in this regard is the problem of interference by some governments with the observance of the code of ethics in scientific research. As regards the professional code of ethics, the point has been rightly made that it is beneficial to have a worldwide code but that the issue is a delicate one because of the origin and locus of most ISPAs in the developed (mostly Western) world.

Carr's chapter on the impact of teachers' associations complements Angell's comprehensive treatment of ISPAs. Its focus on the impact of just one type of ISPA and the historical perspective given with a refreshing personal touch add a significant dimension to our general subject. But the study generally limits itself to the impact of teachers' organizations on the UN system and thus pays little attention to interactions with non-UN intergovernmental organizations, be they functional or regional. Perhaps even more serious is the omission of a discussion of the interaction among the major international teachers' organizations, including the religious ones, and with other international NGOs on the basis of common issues. The discussion of the role of the WCOTP in the adoption of the UNESCO recommendations on the status of teachers shows the dual function of ISPAs, but without clearly pointing out the relationship between their public-oriented and self-interested inclinations.

The initiative of the WCOTP in the early 1960s in promoting the idea of a UN University was significant, but it should also be pointed out that the idea was later revived by Secretary-General U Thant. The UN University that was finally established in 1973 was quite different from the one advocated by the WCOTP in that it comprised a network of research institutions with an administrative core in Tokyo, without a student body.

The Scoble-Wiseberg study on the impact of human rights groups makes a helpful distinction between the violators and promoters of human rights. Of particular interest is the differentiation between the positive and negative roles of governments. In addition, one may question the occasionally sweeping generalizations made: for example, that the targets of skyjackings are from "the most affluent and powerful strata of society" and that because of this governments are expected to respond; and that the prime impact of IGOs has been in the volume of rhetoric and paper (i.e., resulting from conventions and covenants), which is grossly misleading and is somewhat contradicted by the discussion on the importance of international law-making in human rights.

The similarities and differences between human rights groups and ISPAs described are analytically helpful, but the category of human rights NGOs is perhaps too narrow and appears to exclude church-affiliated NGOs like the World Council of Churches which are fundamentally concerned with human rights issues. It should also be acknowledged that some ISPAs are already deeply concerned with the promotion of human rights as we have seen from Carr's chapter on teachers' organizations. Moreover, the confederal system of most ISPAs need not necessarily inhibit them from involvement in sensitive human rights issues: the diversity of perspectives often creates an interaction which may set a trend towards social justice everywhere.

Scoble and Wiseberg have also significantly contributed to the conceptualization and assessment of NGO impact both in terms of success in influencing legislation and in mobilizing a wider public in support of action for human rights. It is clear from the analysis that NGOs have been more influential in the former sense, though the legislative tasks ahead are still quite formidable. As regards public mobilization, a claim is made that "NGOs have had a cumulative impact on American opinion elites and decision makers," but the thrust of the argument does not seem to support this conclusion.

Finally, Scoble and Wiseberg have proposed a sound and generally realistic strategy for NGOs: (i) coalition-building among human rights NGOs, trade unions, church groups, and ISPAs; (ii) further campaigns in specific crisis situations; and (iii) strengthening institutions for human rights. With regard to the third, more effort for the creation of a UN High Commissioner on Human Rights may eventually produce positive results, if only indirectly in the form of acquiring more authority for the intergovernmental bodies on human rights. The new United States policy of linking bilateral foreign aid not only with economic and social needs but also with performance in the field of human rights is promising as long as: (i) too many exceptions are not made to render it arbitrary; (ii) it is directed at the type of aid, particularly military assistance, which tends to help perpetuate repressive ruling elites; (iii) it is not extended to the complex field of multilateral aid where unilateral "sanctions" are likely to be counterproductive.

COMMENTARY

Harold Taylor

My first comment on the Angell and Carr chapters is an expression of gratitude for the interesting information they have provided. My next comments have to do with two points made by Robert Angell in his original study: his distinction between the relative ease on the part of associations of natural scientists as against social scientists in treating the world as a unified system, and the relative speed with which nongovernmental associations can act on specific issues as against the more lumbering governmental organizations.

Angell points out that the natural scientists, geophysicists, oceanographers, meteorologists, and ecologists, for example, treat the world as a unity in their everyday work, and although social scientists do speak of world systems of politics, economics, and social order, "there is so much conflict among nations that few social scientists think of the world as an actually integrated whole."

This may be due in part to the fact that the natural scientists in general were already organized internationally and intellectually long before the social scientists had even begun to organize their own disciplines. It is only in recent years that the cultural parochialism inherent in social science research by reason of limitations on the universality of its data collection has been at least partially overcome by cooperative international research projects. The conventional separation of the natural from the social sciences has also been modified in recent years, as Angell implies, by the necessity for interdisciplinary research in all scientific fields, natural and social, in order to come to grips with world problems.

I do not believe that the existence of conflicts among nations is the reason that so few social scientists think of the world as an integrated whole. Internation conflict is, of course, real, and scientists of all kinds have been assigned by their governments to help their side in the conflicts, in projects ranging from developing new weapons systems to counter insurgency programs. But it seems to me that the reason the social sciences are less internationalized than the natural is that, along with the humanities, they have been taught in the world's schools and colleges that the world is not integrated but is instead made up of nation-states, some of them on "our" side, the rest on the other, including the "nonaligned."

The strain within the international social science community to which Angell refers in his remarks on the cultural heritage of colonialism does exist. Part of the strain is due to a mixture of academic ambition and a certain degree of arrogance on the part of American and other Western researchers who have used their research grants and privileged university status to enter foreign cultures, collect data on topics of their choice, and bundle it up to take back home for use in publications about the cultures they have studied. Years of this kind of academic habit in India understandably turned the Indian government against the whole idea of U.S.-financed research on the problems of Indian society, and led the government to introduce serious restrictions on numbers and fields of study, even before the introduction of Mrs. Gandhi's authoritarian policies.

One good answer to the problem of strain among social scientists is to internationalize fully the research initiated by Western scholars, following the style of the International Federation of Institutes of Advanced Study and the United Nations University. In the latter case, for example, in dealing with the construction of an action-oriented research project on world hunger, the UN University has put together a team of twenty experts whose fields range from nutrition to engineering with connections to the Central Food Technological Research Institute in Mysore, India, and the Institute for Nutrition of Central America and Panama, in Guatemala City, Guatemala.

The question of how advanced in methodology or sophistication the individual scientist is, or in what country the scientist is recognized becomes irrelevant, since the project demands a pooling of research talent and a fair supply of intellectual energy and imagination simply to study problems of global importance. In carrying out the research, each scientist learns from every other, and although qualitative differences in talent and ability of course remain, the differences cannot be put down to the sins of colonialism or the fat research budgets of the United States, or even to the CIA, since the U.S. has given nothing to the United Nations University and the scientists are for all practical purposes denationalized by the nature of the project and its methodology.

Other evidence of strain in the international social science community can be seen in the suspicion among scholars from other countries of the integrity of American research scientists after the Pike report containing revelations of involvement by American academics in work of the U.S. Central Intelligence Agency. The general suspicion had been around for a long time, and the Pike report confirmed what friendly colleagues feared and what hostile critics claimed. Whatever contribution American scientists and intellectuals can make to the welfare of the world community, it must be made outside the framework of paramilitary projects and counter-intelligence programs. It is once more the issue of scholarly integrity, and in affirming that integrity the entire American academic establishment has a prime responsibility. Consider, for example, the preeminence of American technology in the field of satellite scanning and the ability of the scanners to put the whole world under a microscope. Unless American and international influence is exerted for the use of the scanning data for unified and peaceful purposes, national and international conflicts will convert the scanning data into instruments for competition in the discovery of mineral and oil deposits, rather than the detection and prevention of impending famine.

There are also ways of overcoming the deficit in financial support of scientific development in the lesser developed countries by linking U.S. university research projects and postdoctoral studies with work already going on in the universities and research institutes of the Third World. Although there already exists a considerable amount of cooperation between American universities and those in other countries, through which American funds are internationalized in exchange programs and subsidies for cooperative research and research training, much more remains to be accomplished in the creation of networks of universities on a world scale, both inside and outside the existing world bodies which make such cooperative networks possible.

As Robert Angell points out, the International Council of Scientific Unions, with 17 member unions and a variety of internal committees, associations, and commissions, is in unusually good position to develop new networks and arrange for new forms of coordination on a global scale. The General Assembly of ICSU, at its 1976 meeting in Washington, issued an appeal to members of the world scientific community to "rethink their roles and the roles of scientific institutions" in dealing with global problems of nuclear war, world hunger, and the management of world resources. The appeal was specific in its recommendation for action in creating new institutions for developing anticipatory decisions in matters of world policy, as well as creating a climate of opinion on behalf of humane solutions to world problems.

The example of a declaration of this kind brings me directly to William Carr's informative essay on the impact of organizations of teachers on international policy, and in particular, the role of such organizations in establishing UNESCO and the United Nations University. Carr's study and his hands-on history of the activities of the U.S. National Education Association in raising questions and finding answers to the problems of world education is encouraging in two major respects. It demonstrates the possibility of further actions by nongovernmental organizations in influencing world policy, and it reaffirms the role of determined individuals in creating organizations devoted to making such policies.

What Carr does not say is that when he describes the early stages in the mobilization of opinion in favor of world cooperation among teachers and educators, and when he reports the actions of the NEA and the World Confederation of Organizations of the Teaching Profession (WCOTP) at crucial points in the history of international education, he is actually talking about what he did as the key figure in the entire enterprise. Carr was the right man in the right place at the right time. If we needed proof for that general statement, we would need go no further than the documentation Carr provides for his colloquy with Virginia Gildersleeve and Archibald MacLeish on May 12, 1945, when Carr redefined the relation between *culture* and *education,* and could find no good reason for excluding the use of education as a major force in the world's struggle for a peaceable international community.

It is now almost impossible to realize that as recently as 1945, American and world opinion was in such a state of ambivalence about how world peace and social justice could be attained that there could be a dispute about, or even a reluctance to accept, the proposition that education was an instrument for social and cultural change. If it were not for people like William Carr, and for William Carr himself, that idea might not have become embedded in the institutional structure of world affairs in the postwar period.

When international organizations of scientists and intellectuals are studied and written about, as in the present conference, they are usually the ones discussed by Robert Angell—the scientists, technologists and scholars of repute. The status of the teacher, who is in fact the ultimate dispenser of knowledge and humane values in organized society, is considerably below that of lawyers and clergymen, and, as is the case in New York City, parallel with the general public services of firemen, police, and sanitation workers.

Since this is a fact of life, it is encouraging to know that, as Carr points out, the task of disseminating the word about the Universal Declaration of Human Rights and carrying out programs of action through which the rights could become real was taken up by the WCOTP and has been so important an element in the work of the Confederation. It is also encouraging for the future of individual and group efforts by people outside of government that the idea of creating a world university was spread within the world community by WCOTP. By the simple device of inviting President Eisenhower to the 1961 Delegate Assembly of WCOTP, and in one way or another, providing him with a speech advocating the establishment of an international university, the idea was given an unusually wide circulation.

I would like to add one or two items of information to those given by Carr by way of illustrating that those of us who are teachers can have some influence in developing support for world institutions. U Thant, during his young manhood in Burma, was a teacher of English and a modest activist in the Burmese student movement. When in 1963 I had started up a small experiment in a world college in which 23 United Nations countries were involved, U Thant took a special interest in what we were doing and lent his support to the experiment. He already had a lively interest in the United Nations School, and that interest became transferred into the conception of a world university which could bring together on a full scale the students from the entire roster of United Nations countries to work with teachers from every part of the world in a curriculum which represented a global approach to the humanities and the social sciences, assuming that the natural sciences were already safe from nationalist bias.

In his report to the General Assembly in 1969, U Thant proposed the establishment of a United Nations University. His attachment to the idea came from his earlier life as a teacher, his experience with the United Nations School, and his continual search for ways in which nationalist antagonisms and alliances could be transformed into international cooperation.

U Thant's initiative resulted in two feasibility studies, the second prepared by UNESCO, a UN Panel of Experts, a Founding Committee, debates in the Economic and Social Council and the General Assembly, and finally a concrete proposal for a United Nations University, debated and approved by the General Assembly in December of 1973. The proposal has now been put into effect and there is in existence a university bearing the name of the United Nations, with world headquarters in Tokyo, a 24-member world council or board of trustees, and three ongoing research projects dealing with problems of world hunger, economic and social development, and the management of world resources.

The existence of the new university stands as a clear illustration of the way in which not only fully organized groups of educators, scientists, and scholars affect governmental and public policy, but of the role of individuals and loosely organized networks of concerned persons in developing a constituency for new world institutions. The United Nations University required for its establishment an overwhelming majority of votes in the General Assembly, and that majority would never have been achieved without the careful planning and coordinating work of nongovernmental organizations and individuals. The work of Andrew Cordier is a case in point. As chairman of the UN Panel of Experts and of the Founding Committee which wrote the university charter, Cordier brought to the task his years of service in the UN Secretariat and his skill as an educational administrator at Columbia University and its School of International Affairs. It is entirely possible that without Cordier's patient and knowledgeable leadership in shepherding the idea through the controversies and debates of the years from 1969 to 1973, debate which ranged from outright opposition by the Soviet Union to the negative attitudes of the Nixon Administration,[1] the university might never have come into being.

In illustration of the contribution of loosely organized groups of volunteers, we can cite the case of the World Academy of Art and Science, with headquarters in Israel, which drew up plans for a world university and supported the idea through international meetings and research projects coordinated by Harold Lasswell. We can also cite the example of Universities and the Quest for Peace, organized with financial support from Joseph Weingarten, a retired department store owner from Houston,

Texas, and administered by Raga Elim, a political scientist originally from Egypt.

Through the organization of regional conferences during the 1960s in Latin America, Africa, and Europe, Elim and his organization prepared a World Congress, held in collaboration with the University of Vienna in Vienna in 1969, where the President of Austria, with more than three hundred rectors, vice-chancellors, university presidents, UN and UNESCO representatives and scholars from 60 countries, joined together to develop plans for a world university. The publications which resulted from the Congress had wide distribution throughout the international university community and occupied an important place in creating the necessary constituency for the idea of a university sponsored by the United Nations.

The role of the ISPAs and the NGOs continues to be crucial in building that constituency in support of the world's intellectual unity. To this end, these organizations function in three essential ways: (1) by dealing internationally with substantive research and policy issues in the fields in which they are organized; (2) by increasing world support for the idea of using the organized intelligence to solve the problems of human survival and social progress; and (3) by bringing more and more members of the world community into touch with each other in terms of their common intellectual interests and their common concern for the development of a new world order.

NOTE

1. It is an irony among other ironies of the Nixon Administration that, in spite of an official negative policy, one of the strongest proponents of the UN University idea was Glenn Olds, who served in the Nixon Administration as U.S. Ambassador to the UN Economic and Social Council from 1968 to 1971. Among other activities, Olds conferred frequently with U Thant on strategy for developing UN acceptance of the proposal, and organized a series of meetings with a cross-section of UN delegates to arouse support for the idea.

AUTHOR INDEX

Aggarwal, Lalit, 8
Allen, Robert, 163
Amano, A., 105
Andemichael, Berhanykun, 234, 297
Andrews, Fanny Fern, 256
Angell, Robert C., 19, 21, 49, 67, 133, 150, 237, 239, 254, 301, 303

Bachrach, P. S., 292
Bacon, Francis, 11, 24
Baker, F.W.G., 150
Ball, R. J., 92
Beijer, G., 239, 254
Berg, L. L., 292, 278, 283
Berkner, Lloyd, 141
Berner, R., 105
Bernstein, Edward, 81
Bertotti, B., 195
Bjornholm, S., 193
Blau, P. M., 60, 67, 193
Bloomfield, Arthur I., 8
Born, Max, 202
Bragg, Lawrence, 246
Brenner, M., 37, 38, 43, 46
Bridges, D. C., 241, 254
Bridgman, P. W., 202
Bronk, Detlev, 208
Brown, H., 24
Brownlie, I., 273, 292
Brunner, Karl, 87
Bruun, Anton, 140, 141, 144
Buergenthal, T., 273, 283, 292
Bullard, E., 188
Burhop, E.H.S., 176, 200, 203
Burt, J. C., 38, 39, 47
Burton, J. W., 16, 24

Cairncross, Alec, 122
Calder, N., 187
Campbell, D. T., 285, 292
Capurro, Luis, 141
Carr, William G., 19, 21, 255, 256, 257, 271, 299, 300, 301, 303
Cater, Douglas, 227
Cavers, D. F., 188
Cellio, Nello, 219
Chapman, Wilbert H., 143
Chesshire, J. H., 195
Clark, E., 150
Clark, P., 105
Cleveland, H., 107, 120
Cooper, Richard, 80
Cordier, Andrew, 304
Crane, Diana, 15, 17, 20, 21, 24, 29, 69, 70, 71, 74
Crawford, Albert G., 8

Dahl, Robert A., 292
Daysh, Zena, 162
Deacon, George, 140, 141, 144, 150
De Onis, J., 275, 292
De Solla-Price, D. J., 75, 76
Devadas, R. P., 195
Deville, G., 254
Dow, J.C.R., 122
Dubos, R., 160, 173
Duvall, Robert F., 8

Eaton, Cyrus, 176, 227, 231
Echeverria, 219, 220
Eden, M., 188
Einstein, Albert, 23, 176, 202, 227, 267
Eisenhower, D. D., 259, 272
Elim, Raga, 308
Emminger, Otmar, 83
Ennis, J., 247, 254
Erikson, B., 199
Evan, Raima, 8
Evan, W. M., 15, 16, 24, 30, 32, 33, 34, 39, 47, 49, 50, 67, 154, 169, 203, 245, 254
Evang, K., 192
Ewald, P. P., 246, 254

Farmelo, Laura B., 8
Feld, Bernard, 175
Feld, W. J., 49, 67
Fellner, William, 80, 122, 126
Fenn, Wallace O., 243, 247, 254
Feodorov, E. K., 182, 197
Feraru, Anne Thompson, 18, 20, 21, 36, 44, 47, 153, 169, 225, 227, 229
Forrester, J., 209
Forsythe, D. P., 277, 292
Fowler, T., 13, 15, 24
Frank, W., 185, 188
Frey, Frederick W., 8, 225
Friedman, Milton, 128
Frisch, Ragnar, 128
Frutkin, Arnold, 136, 150, 230, 234
Fulbright, J. William, 260

Gabor, Dennis, 215
Galtung, J., 188
Gana, J., 105
Gilbert, Milton, 122

Gildersleeve, Virginia, 261, 304
Glick, C. E., 40, 43, 47
Gordon, R. A., 92
Green, James Frederick, 294
Gvishiani, Jermen, 207

Haberler, Gottfried, 80
Halm, George N., 90, 122, 126
Hardin, G., 13, 14, 24
Harrod, Roy, 80
Hayek, F., 128
Hela, Ilmo, 147
Hernandez-Cata, E., 105
Hickman, Berg G., 92, 105
Higinbotham, W. A., 185
Hill, M. N., 141
Hofstadter, 291, 292
Howe, H., 105
Hughes, Barry B., 18, 21, 107, 120, 127
Humphrey, G., 151
Huntington, S. P., 275, 292
Hurwicz, Leonid, 127
Hymer, S., 274, 292

Infeld, Leopold, 202
Inkeles, A., 16, 24
Iselin, Columbus, 141, 143, 151

Jacobsen, L. R., 105
Johansen, Robert C., 24
Johnson, Harry G., 18, 21, 39, 79, 121, 124, 125, 127, 128, 129
Johnson, J., 151
Johnson, K., 95, 98, 106
Joliot-Curie, Frederic, 202
Jordan, Robert S., 289
Judge, A.J.N., 34, 42, 47

Kende, I., 196
Kenen, Peter, 80
Keohane, R. O., 16, 25, 134, 149, 151, 168, 169, 170
Keynes, 128, 129
Kind, B., 188
King, Alexander, 14, 18, 19, 21, 24, 205, 225, 227, 228, 231, 232
Kissinger, Henry, 122
Klein, Lawrence, 18, 21, 91, 95, 98, 106, 127, 130
Knauss, J., 151
Kort, Vladimir, 144
Kraut, H., 195
Kreisky, Bruno, 219, 220
Kriesberg, Louis, 17, 20, 21, 39, 47, 49, 54, 57, 60, 63, 67, 69, 70, 73, 239, 254
Kristensen, Thorkil, 207
Kurihara, E., 105
Kurth, James R., 69
Kutner, L., 278, 292
Kwack, S., 105

Lador-Lederer, J. J., 133, 151
Lakey, G., 65, 68
Lakoff, S. A., 12, 25
LaLonde, Jean-Louis, 162
Lasswell, Harold, 305
Laszlo, Ervin, 218
Lau, L. J., 105
Lederberg, J., 151
Levinthal, E. C., 151
Lucas, Cyril, 143
Lund, T., 254
Lyons, G. M., 248, 254

Machlup, Fritz, 80, 81, 82, 90, 123, 126
MacLeish, Archibald, 261, 304
Marx, Karl, 128
Mead, Margaret, 161, 164
Meadows, Dennis H., 19, 25, 209, 213
Meltzer, Allan, 87
Merton, Robert K., 16, 25
Mesarovic, Mihajlo D., 18, 21, 25, 107, 127, 215, 217
Miles, Edward, 18, 133, 151, 225, 228, 230, 231
Mill, John Stuart, 211
Miller, J.D.B., 193
Mlinar, Z., 76
Moch, J., 183
Morse, E. C., 76
Muller, H. J., 202
Mundell, Robert A., 79, 80, 86, 87, 88, 90, 121, 123, 124
Mundt, Karl, 260

Nau, H. R., 34, 47
Netter, F., 203
Newton, Isaac, 15
Nixon, J. W., 254
Nobay, A. R., 90
Nossiter, Bernard D., 291, 292
Nye, J. S., Jr., 16, 25, 134, 149, 151, 168, 169, 170

Odishaw, H., 151
Olds, Glenn, 305
Olgaard, P. L., 192
Ozbekhan, Hassan, 208

Palme, Olof, 219

Pannikar, N. K., 141
Parsons, Talcott, 196
Pauling, Linus, 202, 227
Peccei, Aurelio, 207, 216, 217, 227
Pentz, Michael J., 18, 21, 175, 225
Perlmutter, H. V., 274, 292
Perrin, F., 183
Pestel, E., 25, 215, 217
Peterson, N., 194
Plato, 12
Postma, H., 147
Powell, C. F., 176, 202
Prawitz, J., 194
Proudhon, P. J., 128

Rabinowitch, Eugene, 196, 197, 198, 199, 203, 234
Ranuzzi, P., 106
Reichard, John F., 8
Revelle, Roger, 133, 140, 143, 144, 227, 231
Rhomberg, Rudolph, 92
Roosa, Robert, 129
Roosevelt, C., 32, 36, 40, 47
Rosello, P., 256, 272
Rotblat, Josef, 175, 197, 202, 203
Roy, M., 136, 151
Rueff, Jacques, 80
Russell, Bertrand, 23, 176, 202, 227, 231, 267
Russell, Robert W., 121

Sagan, C., 137, 151
Sakamoto, Y., 187, 194
Samuelson, L., 105
Schaefer, Milner B., 143
Schaerf, C., 193
Schmidhauser, J. S., 278, 283, 292
Schoenherr, R., 60, 67
Schroeder-Gudehus, B., 23, 25
Scoble, Harry M., 19, 20, 21, 273, 278, 284, 290, 292, 293, 294, 296, 299, 300
Scully, M. G., 287, 292
Searle, Graham, 165
Senghor, 219
Servan-Schreiber, J. J., 216
Skjelsbaek, K., 16, 25, 30, 47, 49, 68, 133, 151
Skolnikoff, E. B., 137, 151
Slovo, Gillian, 18, 21, 175, 225
Smith, Adam, 128
Smith, A. K., 203
Smith, Ph. B., 187
Smoker, P., 32, 33, 47, 133, 151
Sneath, P.H.A., 152
Sohn, L. B., 273, 292
Speeckaert, G. P., 42, 47
Staley, E., 187, 195
Stanley, J. C., 285, 292
Strong, Maurice, 160, 164
Sugawara, K., 141
Sverdrup, Harold, 140
Swoboda, A. K., 86, 90

Taft, Robert A., 260
Tamm, I. E., 197
Taylor, Harold, 303
Ter Haar, D., 15, 25
Teune, Henry, 73, 76
Thant, U, 299, 305
Thiemann, 227
Tinbergen, Jan, 128, 217
Todd, J. E., 36, 47
Triffin, Robert, 80, 126
Trudeau, P., 219
Truman, David, 227

Van der Ueyel, 219
Van Dyke, V., 284, 293
Van Lennep, Jonkheer, 82
Van Putten, J. G., 162, 163

Waelbroeck, J., 98, 106
Wallace, M., 291, 292
Walsh, J., 24, 25
Ward, Barbara, 160, 164, 173
Warren, R. L., 43, 47
Weinberg, A. M., 199
Weisman, John, 140
Weiss, Paul, 208
Weissbrodt, D., 293
Wenk, E., 138, 152
Whaley, Randall M., 8
White, L., 133, 152
Wiesner, J., 182
Wilson, Carroll, 207
Winings, T. A., 34, 47
Wiseberg, Laurie S., 19, 20, 21, 273, 278, 284, 290, 292, 293, 294, 296, 299, 300
Wooster, W. S., 37, 43, 47, 140, 145, 146, 147, 152, 203, 227

Yukawa, H., 202

Zimmern, Alfred, 263, 267
Zweig, M., 267, 272

SUBJECT INDEX

Academia dei Lincei, 207
academic economists, cooptation by policy makers, 125
Academy, The (Plato), 15
"access," 227
"acronymia," 225
actors, transgovernmental, 16; transnational, 16, 17
advisers, scientific, to governments, 200
AFL-CIO, political clout of, 280
alliances, tripartite, 227
American Association for the Advancement of Science, 287
American Bankers Association, political clout of, 280
American Bar Association, 278, 284
American Mathematical Society, 287
American Medical Association, political clout of, 280
Amnesty International, 277, 279, 295
analysis techniques, of systems, 108
analytic model, systems, 18
antislavery, 277
APT systems, 110, 111, 113, 114
Assessment of Policies Tool, 110
associations, international scientific and professional (ISPAs), 8, 37; politicization of, 280
associations, professional, 237
associations, scientific, 237
associations, transnational, 8
associations, voluntary, 32
Astronomical Union, International, 252
Atmospheric Research Program, Global, 252

balance of payments, 115
bankers, central, 87
Bellagio Group, 79, 80, 121, 128, 129
Berlin Wall, 265
bigness, value of, 279
Bill of Rights, International, 295
Bricker Amendment, 283
budgets, of ISPAs, 58
Burgenstock Group, 122, 125

calculus, of "National interests," 7
Campbell's design, quasi-experimental, 285
canons, of scientific inquiry, 7
capital transfer, 99
cave, idols of, 13
CMEA, 93
centers, and peripheries, 75
central bank, international, 83, 124
channels, communication, 232
channels, institutionalization of, 188
Club of Rome, 19, 107, 109, 111; early history of, 206; member reaction of, 213; ministerial meetings of, 218; national associations for, 220; organization of, 223; policy impact of, 205-224; political influence of, 231
coalitions, ad hoc, of ISPAs, 168; formation of, 20; informal, 286, 287; transgovernmental, 168, 169
Committee on Space Research (COSPAR), 18; political influence of, 135-139, 148, 149
commodity market models, 100
commons, tragedy of, 13, 14
communality, 16
community, cultural, 14; international scientific, 15; scientific, 14, 23; territorially based, 14
computer technology, 128
Conference on Law of the Sea, 232, 234; response of SCOR to, 146
conflicts, domestic, 274
consciousness, transnational, 23
conservation, law of, 93
consultative relationships, network analysis of, 44
consumer theory, 104
Convention Barring the Development, Manufacture and Stockpiling of Biological Weapons, 19
cooperation, international scientific, 198
cooptation, 125, 241
copyright law, international, 282
corporations, multinational, 70, 274
cosmopolitanism, 85, 126
COSPAR, 18, 229; political significance of, 135, 139
counterconferences, 229
"crawling peg," of market rate adjustments, 85
cycle, business, 92, 98

cycles, countermeasures, 99, 101

decision-making, global, 7; international, 8; political, 7
decolonization, 75
development, global, 76, 111
dialogue, man-computer, 216
differentiation, hierarchical, 58
dimensions, of international system, 17
disarmament negotiations, 185, 191, 198; pressure groups on, 188
disinterestedness, 16
Displaced Persons (Nurses) Register, 244
dissidents, ideological, 22
Dumbarton Oaks, 260

"early-warning" system, 288
Earthwatch, 162, 167
Econometric Society, 128
econometrics, 84, 130
economic development, 116
economic order, international, 217; new, 217; reshaping, 217
economic stabilization, 92
economic system, world, 107
economics, mathematical, 84
economists, networks of, 18, 79
economy, world, 91, 93, 99, 103; disarray in, 7
ECOSOC, 43, 244, 251, 252
Education, International Bureau of, 255
elites, national, 280; political, 280; ruling, 300; scientific, 22
embargo, of art, 99
energy model, 115, 116
enterprises, multinational, 16, 17; regional, 16
Environment, Scientific Committee on the Problems of, 252
ethics, code of, 298, 299
European Common Market, 216, 211
European Coordination Centre for Research and Documentation in the Social Sciences, 252
European Economic Commission, 212
exchange rates, 99; alternatives of, 98; final, 88; flexibility of, 122, 123, 125; floating, 81, 84, 85, 88, 98; revaluation of, 98; simulations of, 91
experiments, in space, 136

Exxon, 70

fiscal policies, of U.S., 98
Florence Nightingale International Foundation, 243, 244
Ford Foundation, 268
formation, of coalitions, 20
freedom, of association, 22; of expression, 22; of research, 22
frequency allocation for radio astronomy and space science, interunion commission on, 252

geodesy, international union of, 251
geophysics, international union of, 251
global communities, 49
global development, 76
global humanism, 8, 23
global modeling, 107-120; multilevel, 109
global organizations, 65
global problems, solution for, 7
global social integration, 251; promoted by ISPAs, 237-254
global society, 59, 239; emergence of, 14; values of, 12
Gnomes of Zurich, 212
goals, transnational, 39
gold exchange standard, 81, 84, 85. See also exchange rates
goodwill, transnational, 247
groups, informal, 224
growth, exponential, of INGOs, 75, 76; GDP, world, 95; limits to, 108, 109; negative, 212; organic, 109; zero, 213
growth ethic, 108

Habitat, 153
Helsinki Conference on Security and Cooperation in Europe, 297
hierarchical differentiation, 58
Hiroshima, and *crise de conscience,* 175
"hot line," 114
human rights, 280; constituency for, 284; international league for, 279; UN commission on, 290, 291
human rights legislation, codification of, 283
human rights organizations, 19, 273
human rights treaties, ratification of, 284

human rights violations, 274; governmental inertia and indifference to, 276
humanism, global, 8
humanist appeal, global, 23
hunger and starvation, 7

ICSUs, 18, 36, 37, 43, 44, 70, 133, 134, 138, 140, 160, 202, 229, 230, 303
ideological dissidents, 22
"idols," of cave, marketplace, theater, tribe, Bacon's concept of, 12, 13, 23
IFIAs, 221
IGOs, 17, 35, 36, 38, 43, 69, 70, 251
IGO–INGO relations, politicization of, 251
ILO, 242, 252
Industrial Revolution, 208
inflation, world, 83, 88, 95, 96
influence, channels of, 20, 126
INGOs, 29, 36, 38, 44, 73, 133, 153, 227, 253; at Habitat, 164; budget of, 58; categories of, 50; consultative status of, 58, 166; emergence of, 225; meetings of, 58; transnational impact of, 58
inquiry, canons, of scientific, 7
integration, intersocietal and intercultural, definition of, 239, 240; global, 244, 245, 246, 251; influence of, 253; nonlinear model of, 202
integration approach, to and of ISPAs, 29
integrity, scholarly, 302
interdependence, of nations, 107
interest, objectives of, 157
interest group approach, of ISPAs, 29
interest groups, issue/phase distinctions in, 227; ISPAs as, 34; objectives of, 157
interests, aggregation and articulation of, 230; nongovernmental, 167
intergovernmental conferences, 154
International Assembly of Non-Governmental Organizations, 161
International Bill of Rights, 295
International Committee of the Red Cross, 238, 250
International Conference of Midwives, 250
international conferences, political impact of ISPAs at, 166
International Council of Nurses, 249
International Council of Scientific Unions (ICSUs), 15, 18, 22, 70, 133, 134, 229, 230. See also ICSUs
International Dental Foundation, 247, 250
International Federation of Institutes of Advanced Study, 221, 302
International Geophysical Year, 230
international governmental organizations, 29, 69. See also IGOs
International Hospital Association, 238
international integration, and ISPAs, 33
international interactions, cultural dimensions of, 228
International Labor Office, 242
international labor organizations, 294
international law, 276
International League for Human Rights, 277
International League for the Rights of Man, 277
International Library Association, 238
international liquidity, 83
International Monetary Fund, 79, 80
international monetary reforms, 79-90
international nongovernmental organizations, 29, 49, 73, 133; activity and influence, 271; consultative status with UN Commission on Human Rights, 290-296. See also INGOs
international relations, impact of ISPAs on, 70
International Social Science Council, 15, 22
International Statistical Institute, 240, 247
international system(s), 69, 71, 74; influence of ISPAs on, 29; perspective on, 76
International Telecommunications Union, 243
International Telegraph Union, 243
International Union of Crystallography, 246
International Union of Physiological Sciences, 247
International University, 267

International Women's Conference, 153
invisible college, international, 18, 37
ISPAs, 8, 19, 22, 43, 44, 49, 71, 225, 230, 231, 252; age of, 59; age of membership of, 61; alternative models of, 29; and international integration, 32; approaches to, 32; as interest groups, 34, 39; as transnational actors, 16; budgets of, 58; characteristics of, 54; contributions to IGOs of, 249; demography of, 30; exponential growth of, 75; forms of, 49; functions of, 49; goals of, 59, 61; governmental contact of, 157; impact upon transnational relations of, 70; influence on international system of, 18, 20; international ties among, 64; leaders of, 228; meetings of, 58; membership, type of, 59, 61; network interactions of, 20; objectives of, at UN conferences, 155; organizational aspects of, 59; participants of, at UN conferences, 154; political impact of, at international conferences, 166; relationship with other international organizations of, 63; role in international concerns of, 229; staff of, 58, 59; 61; structural attributes of, 17; technical and scientific expertise of, 167; varieties of, 59
ISPAs and NGOs, difference between, 278
ISSC, 252

jurists, international commission of, 277, 278, 279
justice, attainment of, 64

Keynesian models, 130; anti-, 130
knowledge and power, dilemmas of, 11-25

Lasswellian values, 280
Latin American Social Science Council, 249
law, implementation and adjudication impact of, 281; international, 276; of the sea, 140, 143, 146, 153, 232
"Le World Problematique," 206, 218, 223, 226
leadership quality, as catalyst, 231
League of Nations, 252; covenant of, 256
League of Red Cross Societies, 238
legal scholars, international, 281
limits to growth, 108, 104
Limits to Growth, 212, 213, 214, 216
LINK Project, 18, algorithm of, 103, 104; applications of, 94, 98; model of, 127, 128, 129; policy uses of, 98; simulation of, 98, 99
linkages, international interorganizational, 20, 61
liquidity, international, 83, 124
lobbying, objectives of, 157
loyalties, transnational, 23
Lunar Society, 208

macroeconomic policy, Keynesian approach to, 129
margin, of fluctuation of exchange rates, 85
marine science, 148; international system for, 145
marketplace, "idols" of, 13
meetings, of ISPAs, 58
meteorology, international, 141, 251
methodology, strategic, network analysis as, 40
Midwives, International Conference of, 250
mind, "phantoms of the," 13
mini-states, 17
MNCs, 70, 274
modeling, world multilevel, 109
model(s), of commodity market, 100; econometric, 92, 93; forecast of, 94, 95; macroeconomic, 91, 93; naive, of ISPAs, 121, 123, 124; simulation of, 94; world, 93
monetary system, international, 18, 86, 89, 121, 122, 123; reform of, 121, 122, 123
multinational corporations, 70
multitrait-multimethod approach, to impact, 286

NASA, 136, 137
nation-states, 15, 17, 67, 276
National Academy of Science, 287

National Council of Churches of Christ in the USA, 290
National Education Association, 259, 263, 303
"national interest" orientation, of governmental representatives, 167
national interests, calculus of, 7; official perceptions of, 147, 148
national nongovernmental organizations (NNGOs), 29
nationalism, 85
NGOs, 153, 154; and ISPAs, difference between, 278; coalition building of human rights in, 286, 287, 288; influence of, 284; national, 163
network(s), governmental, nongovernmental, 145; informal, of scientists and professionals, 253; informal international, 20; interlocking, 87; international, 125; of consultative relationships, 41; of economists, 18, 79, 121; of information and knowledge exchange, 114; overlapping, 87
network analysis, 29, 74; as a strategic methodology, 40; of consultative relationships, 44
network systems, academic, 87
New Atlantis (Bacon), 11
New International Economic Order, 222
Non-Proliferation Treaty, 197
"North-South negotiations," 234
notables, critical roles of, 227
Novum Organum (Bacon), 12
Nuclear Non-Proliferation Treaty, 19
nuclear weapons, 7, 176; development and testing of, 177, 183
nuclear-free zones, 177, 197

ocean, freedom of scientific research in, 147
oceanography, 141, 142, 144
oil embargo, 99
order, international, humanization of, 277
organization(s), for human rights, 273; global, 65; intergovernmental, 223, 224, 277; international governmental, 17; international nongovernmental, 73, 226; international, proliferation of, 16; nongovernmental, 277; size of, 62; vertical differentiation of, 62
Organization For Economic Cooperation and Development, 239
organizational aspects, of ISPAs, 59
organizational characteristics, 54
organizational density, of world, 58
overpopulation, 7

parallelism, structural, 226
parochialism, 228; cultural, 303
P.E.N. International, 277, 278, 279
Penal code, of West Germany, 276
peripheries, and centers, 75
"phantoms of the mind," 13
Planned Parenthood, International Federation of, 250
pluralistic world system, 107
political system, international, 169
pollution, environmental, 7
Population Conference, World, 153
Population, International Union for the Scientific Study of, 242
Principle of Free Circulation of Scientists, 24
Principle of Universality, 24
problems, global, 1, 8
processes, of science, 7; of decision-making, 7
production, world, 96
products, of science, 7; of decision-making, 7
Project LINK, 18, 91-106; origins, 91; overview, 93. See also LINK
public opinion, world, 280
Pugwash Conference on Science and World Affairs, 19, 23, 64, 176, 177, 178, 179, 183, 185, 187, 189, 196, 198; aims of, 189; attitudes toward, 191; effects of, 190, 196; political significance of, 175-203

radiation units and measurements, international commission on, 250
Radio Science, International Union of, 251, 252
radiological protection, international commission on, 250
rationality, global, 114
recession, world, 99

Red Cross, International Committee of, 277, 279
regimes, totalitarian, 274
relations, consultative, 251, 252
relationships, with IGOs, 36
representation, objectives of, 157
repression, political, 274
Research Group for European Migration Problems, 238, 239, 242
research, international, cooperative, 303
resource potential, 279
resource utilization, 149
resources, nonrenewable, exhaustion of, 7
revolution, scientific, 234
rights, of authors, 282
Royal Society of London, 15
Russell-Einstein Manifesto, 176

SALT I, negotiations and agreements, 177, 197
scenario analysis, 93
science, ethos of, 16; processes and products of, 7; social function of, 175
scientific activity, internationalization of, 74
Scientific Committee on Oceanic Research (SCOR), 18; political influence of, 139-149
scientific communities, 23; transnationals, 22; world, 303
scientific elites, 22
scientific inquiry, canons of, 7
scientific planning, 149; impact of SCOR on, 145
Scientific Unions, International Council of, 251
scientists, attitudes of, 198; powerlessness of, 188; social responsibilities of, 175
SCOR, 18
SDR negotiations, 124
self-interested actors, ISPAs as, 29, 39
simulations, of capital transfer, 99; of LINK policy, 99; of models, 94
skepticism, 16
skyjackings, 276
slavery. See antislavery
social research, comparative, 66
social science, international community, 304
Social Science Council, International, 252
society, global, 13
society, nonstate, 15
sociology, internationalization of, 245
space, experiments in, 136; exploration of, 197; committee on, 252
space science, 148
Special Drawing Rights, 83
Sputnik I, 143, 144
"state-centric paradigm," 16
staff, of ISPAs, 58
Standardization, International Organization For, 250
Statistical Society of London, 240
strategies, alternative, 286
structures, orthogonal, 226
subgovernments, 227
submodel, world trade, 115
suffragette movement, 277
superpowers, 27
supranational agencies, 102, 276
system(s), analytic model of, 18; dynamic approach of, 209; international, 17, 74; political, 75
systems analysis techniques, 108
systems theory, general, 109

Teachers, Catholic, World Union of, 265
Teachers' Charter, 260
Teachers, International Organizations of, policy impact, 255-272
teachers, organizations of, 19; international, 263, 270; lack of unity in, 270; need for concerted efforts in, 270; shortage of resources of, 270
Teachers Syndicates, International Federation of, 263, 264, 265
Teaching Profession, World Confederation of Organizations of, 264
technocrats, 39
"technological optimists," 215
telecommunications system, 71
terrorism, 275, 276
Test Ban Treaty, 19, 200
theater, "idols" of, 13
theory, universal, 85
torture, as crime against humanity, 282

tourism, 17
trade, imbalances of, 98
trade, world, 95, 96
trade liberalization, 85
trade matrix equations, 104
trade theory, international, 84
"tragedy of the commons," 13, 14
transgovernmental coalitions, 168
transgovernmental relations, 168
transition, of values, 23
transnational actors, 70, 74, 225, 226
transnational corporations, role of, 217
transnational goals, 39
transnational goodwill, 247
transnational negotiation, 230
transnational strategy, 148, 230
transnational structures, alternative, 65, 66
transnationalism, 34; in scientific community, 22
tribe, "idols" of, 12, 23
Trilateral Commission, 275
Tuberculosis, International Union Against, 251

UNESCO, 19, 22, 35, 140, 202, 212, 232, 242, 246, 251, 252, 256, 258, 265, 294, 297, 299, 303; constitution of, 262, 263
Union of International Associations, 251
United Nations, 143, 225; creation of, 258
United Nations Charter, 258, 259, 262, 297
United Nations Children's Emergency Fund (UNICEF), 265
United Nations Commission on Human Rights, 290
United Nations Conference on Human Settlements, 153
United Nations Conference on the Human Environment, 153; ISPA activities before, at, and after, 159, 160, 161, 162
United Nations Conference on the Law of the Sea, 18, 140, 146, 153
United Nations system, impact of WCOTP on, 265
United Nations University, 267, 299, 302, 303, 305, 306
Universal Declaration of Human Rights, 22, 265, 266, 281, 304
universalism, 16
Universality, Principle of, 24
universities, international, 19
University, The International, 267
University, United Nations, creation by General Assembly, 267

values, Lasswellian, 280; nationalistic, 13, 23; of a "global society," 12; of global humanism, 24
variables, noneconomic, 127
vertical differentiation, in organizations, 62

Water Research, Committee on, 252
WCOTP, 265, 266, 267, 268, 269, 299
West German Penal Code, 276
"whistle-blowers," 2
WHO, 43, 202
WIM, 114
wisdom, conventional, 13
World Academy of Arts and Science, 305
World Bank, 100
world community, 66, 67, 302. See also world society, global society
World Confederation of Organizations of the Teaching Profession, 303
World Council of Churches, 238
world culture, development of, 64
World Federation of Mental Health, 244
World Hospital Federation, 244
world integrated model, 110; structures, diagrams of, 114, 116
World Medical Association, 246, 298
world model, disaggregated, 215
World Population Conference, 153
world power structures, 218
world problems, 153
world recovery, 99
world society, 49, 60, 64. See also global society, world community
world systems, 113, 206; alternative futures of, 113, 114; state of, 113
world trade model, 120
world trade submodel, 115
world university, 306

ABOUT THE AUTHORS

ROBERT C. ANGELL is Emeritus Professor of Sociology, University of Michigan. His major publications include *A Study in Undergraduate Adjustment* (1930), *The Family Encounters the Depression* (1936); *The Integration of American Society: A Study of Groups and Institutions* (1941), *The Moral Integration of American Cities* (1951), *Free Society and Moral Crisis* (1958), *Peace on the March: Transnational Participation* (1969), and *The Quest for World Order* (1979).

BERHANYKUN ANDEMICHAEL is Assistant Director of Research of UNITAR and Special Assistant to the Executive Director of UNITAR. His current research interests are in African international relations, disarmament negotiations, regional and global organizations, and nongovernmental organizations. He is the author of *The OAU and the UN: Relations Between the Organization of African Unity and the United Nations* (1976), and editor of *Regionalism and the United Nations* (1979).

WILLIAM G. CARR is President of the Sino-American Cultural Society. His current research interests are Chinese-American relations and the evaluation of documentary films for use in international film festivals. His major publications include: *Education for World Citizenship* (1928), *County Unit of School Administration* (1931); *John Swett: Educational Pioneer* (1933), *School Finance* (1933), *International Frontiers in Education* [editor] (1944), *Only By Understanding* (1946), *One World in the Making* (1946), *Collecting My Thoughts* (1980).

DIANA CRANE is Professor of Sociology at the University of Pennsylvania. She is the author of *Invisible Colleges, The Sanctity of Social Life: Physicians' Treatment of Critically Ill Patients,* and numerous journal articles, including "Studies of Science and Technology Policies" in Paul Durbin (ed.) *A Guide to the Culture of Science, Technology and Medicine.* She is currently studying the social factors in the evolution of artistic styles.

WILLIAM M. EVAN is Professor of Sociology and Management at the University of Pennsylvania. He is the author of numerous articles for various professional journals; he is also the author or editor of *Preventing World War III* (1962), *Law and Sociology* (1962), *Organizational Experiments* (1971), *Organization Theory* (1976), *Interorganizational Relations* (1976, 1978), *Frontiers in Organization and Management* (1980), and *The Sociology of Law* (1980). His principal current research interests are organization theory and the sociology of law.

ANNE THOMPSON FERARU is Professor of Political Science at the California State University, Fullerton. Her research interests continue to be focused on the international activities of nongovernmental organizations, and her articles on this subject have appeared in *International Organization, Transnational Associations,* and *ECOFORUM.*

FREDERICK W. FREY is Professor of Political Science at the University of Pennsylvania. His research interests include analysis of power, political socialization and comparative politics. His major publications include *The Turkish Political Elite,* "Survey Research on Comparative Social Change," and "Cross-Cultural Survey

Research in Political Science," in Robert T. Holt and John E. Turner (eds.) *The Methodology of Comparative Research.* He is currently completing a book on *The Analysis of Power.*

JAMES FREDERICK GREEN is Executive Director of the Commission to Study the Organization of Peace. His main field of research interest is human rights. His publications include *The United Nations and Human Rights* (1956); "NGOs," in Abdul Aziz Said (ed.) *Human Rights and World Order* (1978); and "Decision Making on Economic and Social Questions in the United Nations," in Seymour Maxwell Finger and Joseph R. Harbert (eds.) *U.S. Policy in International Institutions: Defining Reasonable Options in an Unreasonable World* (1978).

BARRY B. HUGHES is an Associate Professor at the Graduate School of International Studies, University of Denver. He is the author of *The Domestic Context of American Foreign Policy* and of *World Modeling.* He continues to work with global models and is developing one suitable for classroom use.

LEONID HURWICZ is a Regents' Professor of Economics and a member of the National Academy of Sciences. His current research interests include the design of economic systems and welfare economics problems in "nonclassical" situations. Among his publications are "On Allocations Attainable Through Nash Equilibria," *Journal of Economic Theory* (1979); "Outcome Functions Yielding Walrasian and Lindahl Allocations at Nash Equilibrium Points," *Review of Economic Studies* (1979); and *Studies in Resource Allocation Processes,* edited with Kenneth J. Arrow (1977).

HARRY G. JOHNSON, the late Charles F. Grey Distinguished Service Professor of Economics at the University of Chicago, was the author of numerous publications, including *Money, Trade and Economic Growth* (1962), *The World Economy at the Crossroads* (1965), *Essays in Monetary Economics* (1967), *Macroeconomics and Monetary Theory* (1972), *General Equilibrium Analysis* (1976).

ROBERT S. JORDAN is Dean of the Graduate School and Professor of Political Science at the University of New Orleans. His major publications include *The International Civil Service: Changing Role and Concepts* [with Norman Graham] (1980); *Political Leadership in NATO: A Study in Multinational Diplomacy* (1979), *The World Food Conference and Global Problem-Solving* [with Thomas G. Weiss] (1976), and *Assessing Global Conferences: A Comparative Inquiry* [with Norman Graham and Stephan Haggard, for the United Nations Institute for Training and Research] (1981). He is currently working on a book on Dag Hammarskjold for the Institute of International Studies, University of South Carolina.

ALEXANDER KING is the chairman of the International Federation of Institutes of Advanced Study (IFIAS) and co-founder of the Club of Rome. His main interests are at present science and technological policy, transdisciplinary research promotion and holistic studies. He is the author of numerous papers on physical chemistry, science policy, management and education; and three textbooks on chemistry: *The State of the Planet* (1980); *Science, Technology and the Global Problems* [edited with Maurice Goldsmith] (1979); and *Bioresource Development* [edited with Harlan Cleveland] (1980).

LAWRENCE R. KLEIN, Nobel Laureate in Economics, is Benjamin Franklin Professor of Economics and Finance at the University of Pennsylvania. He is presently involved in continuing work on the world economy, extending the LINK system in various dimensions, enlarging country coverage, expanding commodity detail, lengthening the simulation horizon, explaining capital flows and exchange rates, explicitly modeling large developing countries, and making new applications to contemporary policy debates. Recently, he has been active in introducing modern econometrics to the People's Republic of China. Applications of econometrics to domestic policy in the United States is a continuing area of present activity. His publications include *An Econometric Model of the United States, 1929-52; The Keynesian Revolution; A Textbook of Econometrics; An Introduction to Econometrics; The Wharton Econometric Forecasting Model* (with M. K. Evans); *An Introduction to Econometric Forecasting and Forecasting Models* (with R. M. Young); and "A Decade of Research by Project LINK" (with B. G. Hickman), *ITEMS* (1979).

LOUIS KRIESBERG is Professor of Sociology at Syracuse University. He is currently engaged in research on noncoercive inducements in international conflicts, examining peace initiatives in the context of the U.S.-Soviet and Arab-Israeli conflicts. He is the author of *Mothers in Poverty; The Sociology of Social Conflicts;* and *Social Inequality;* and is the editor of *Social Processes in International Relations* and of the annual series *Research in Social Movements, Conflicts and Change.*

JAMES R. KURTH is Professor of Political Science at Swarthmore College. His current research interests are the impact of industrial interests and technological innovations on foreign policies and international politics. His major publications include "The Political Consequences of the Product Cycle: Industrial History and Political Outcomes," *International Organization* (1979); "Industrial Change and Political Change: A European Perspective," in David Collier (ed.) *The New Authoritarianism in Latin America* (1979); *Testing Theories of Economic Imperialism* [edited with Steven J. Rosen] (1974).

MIHAJLO D. MESAROVIC is a visiting professor at the School of Engineering at MIT. His current research interests are large scale social systems theory and world modeling. His principal publications are *Mathematical Theory of General Systems* (1977) and *Mankind at the Turning Point* [with Edward Pestel] (1974).

EDWARD MILES is Professor of Marine Studies and Public Affairs at the University of Washington. His current research interests include restructuring the world ocean regime; the effects of regional implementation of exclusive economic zones on the management of ocean resources and ocean uses; the management of marine resources and uses in the north Pacific Ocean; and international science and technology policy. Some of his major publications are *The International Administration of Space Exploration and Exploitation* (1971); "Transnationalism in Space: Inner and Outer," *International Organization* (1971); *Law of the Sea: Problems of Implementation* [edited with John Gamble] (1977); and *The Management of Marine Regions: The North Pacific* [with Steven Gibbs et al.] (in press).

MICHAEL J. PENTZ is Dean and Director of Studies in Science in The Open University in Great Britain. His current research interests include studies of global energy problems and strategies and current developments and implications of the nuclear arms race. His publications include research papers on air-cored betatrons for controlled fusion and on microwave spectroscopy in *Nature* and in the *Proceedings of the Physical Society* (1949-1958); and research papers on beam-stacking accelerators and electron storage rings in *Proceedings of International Conference on High Energy Accelerators* (1959-1968).

ROBERT W. RUSSELL is Counsel to the International Finance Subcommittee of the U.S. Senate Banking Committee. His current research interests are international trade and monetary policy. His major publications include "The External Effects of the European Monetary System," in Werner J. Feld (ed.) *Western Europe's Global Reach: Regional Cooperation and Worldwide Interests* (1980); "Evaluating U.S. Policies Toward International Trade and Investment," in Vincent E. McHale (ed.), *Evaluating Transnational Programs in Government and Business* (1980); and "International Monetary Reform: Don't Just Do Something, Stand There," *International Organization* (1977).

HARRY M. SCOBLE is Executive Vice Chairman of Human Rights Internet. His current major research interest is nongovernmental organizations and human rights. His publications include *Ideology and Electoral Action* (1967); "Access to Politics," *International Encyclopedia of the Social Sciences* (1968); and "Human Rights NGOs: Notes Toward Comparative Analysis" [with L. S. Wiseberg], *Human Rights Journal* (1976).

GILLIAN SLOVO is a researcher with Counter Information Services. She is currently working on a study of the arms industry in Britain.

HAROLD TAYLOR is Distinguished Professor of the Social Sciences, City University of New York, the College of Staten Island; and Director of the Center for International Service. He is currently writing a book on the cultural history of the 1950s. His publications include *The World as Teacher* (1970), *Students Without Teachers* (1969); *How to Change Colleges: Notes on Radical Reform* (1971), *A University for the World: The United Nations Plan* (1975), *Art and the Intellect* (1963), and *On Education and Freedom* (1975).

HENRY TEUNE is currently Professor of Political Science at the University of Pennsylvania, and President-elect of the International Studies Association. His recent publications include *The Developmental Logic of Social Systems* [with Mlinar] (1978), and *The Social Ecology of Change* [edited with Mlinar] (1978). In 1980 he co-edited and contributed to Volume 2 of the *International Political Science Review* the article "Recent Changes in Urban Politics." His two recent articles on methodological issues in the social sciences are "Cross-Level Analysis: A Case of Social Inference" (1979) and "Concepts of Evidence in Macro Systems Theories" (1980), both appearing in the journal, *Quantity and Quality*. He is completing a manuscript on "The Future of American Democracy."

LAURIE S. WISEBERG is Executive Director of Human Rights Internet. Her principal current research interests are nongovernmental organizations and human rights. Her publications include "Humanitarian Intervention: Lessons from the Nigerian Civil War," *Human Rights Journal* (1974); "Monitoring Human Rights Violations: The Role of Human Rights NGOs" (with Harry M. Scoble), in Donald P. Kommers and Gilbert D. Loescher (ed.) *Human Rights and American Foreign Policy* (1979); and "Human Rights Protection: Research Agenda" (with David P. Forsythe), *Universal Human Rights* (1979).